Communicatin

This book is a wide-ranging exploration of PR and communication in the construction industry, with a strong emphasis on communications theory, strategy and technique.

The editors begin with an introduction to the UK construction industry and its supply chains, as well as various elements of PR in relation to the construction process. Subsequent chapters provide a strategic overview, practical examples, success stories, case studies and personal perspectives on PR for different parts of the built environment and reputational issues in construction. Chapters include expert advice on communications for architecture, planning, building consultancy, building products and manufacturers, general and specialist contractors, construction technology, infrastructure and communicating sustainability in the built environment. The conclusion looks at the current and upcoming reputational priorities for communicators in construction, as well as the top ten priorities for implementing PR as a strategic management discipline in the industry.

This book is essential reading for all construction PR teams, students studying both for built environment and PR/marketing degrees and CPD courses, and anyone working in the built environment sector who needs to consider PR and marketing as part of their role.

Liz Male MBE is the founder of LMC (Liz Male Consulting Ltd). She has 30 years' experience of marketing communications, corporate reputation and issues-led PR for the property and construction industry in both agency and in-house roles.

Penny Norton has worked in construction PR for 20 years, during which time she has covered all elements of communications for the built environment, from pre-planning consultation through to media relations for completed schemes. Penny is the author of *Public Consultation and Community Involvement in Planning: A Twenty-First Century Guide* and she writes extensively for construction publications.

'This is the ultimate book on construction PR. It combines a broad strategic approach along with some excellent examples of best practice. The diversity of the chapter authors and their background demonstrates the incredible extent and breadth of the sector and such a broad approach will undoubtedly help to equip both those entering the industry and those already working within it with a wide perspective, as well as some fantastic insight.'

Emma Leech, CIPR President 2019

Communicating Construction

Insight, Experience and Best Practice

Edited by Liz Male and Penny Norton
with a foreword by Alastair McCapra

LONDON AND NEW YORK

First published 2021
by Routledge
2 Park Square, Milton Park, Abingdon, Oxon OX14 4RN

and by Routledge
52 Vanderbilt Avenue, New York, NY 10017

Routledge is an imprint of the Taylor & Francis Group, an informa business

© 2021 selection and editorial matter, Liz Male and Penny Norton; individual chapters, the contributors

The right of Liz Male and Penny Norton to be identified as the authors of the editorial material, and of the authors for their individual chapters, has been asserted in accordance with sections 77 and 78 of the Copyright, Designs and Patents Act 1988.

All rights reserved. No part of this book may be reprinted or reproduced or utilised in any form or by any electronic, mechanical, or other means, now known or hereafter invented, including photocopying and recording, or in any information storage or retrieval system, without permission in writing from the publishers.

Trademark notice: Product or corporate names may be trademarks or registered trademarks, and are used only for identification and explanation without intent to infringe.

British Library Cataloguing-in-Publication Data
A catalogue record for this book is available from the British Library

Library of Congress Cataloging-in-Publication Data
Names: Male, Liz, editor. | Norton, Penny, 1973- editor. | McCapra, Alastair, writer of foreword.
Title: Communicating construction: insight, experience and best practice / edited by Liz Male and Penny Norton; with a foreword by Alastair McCapra.
Description: First edition. | Abingdon, Oxon; New York, NY: Routledge, 2021. | Includes bibliographical references and index.
Identifiers: LCCN 2020039440 (print) | LCCN 2020039441 (ebook)
Subjects: LCSH: Construction industry–Communication systems. | Construction industry–Public relations.
Classification: LCC TH215 .C647 2021 (print) | LCC TH215 (ebook) | DDC 624.068/4--dc23
LC record available at https://lccn.loc.gov/2020039440
LC ebook record available at https://lccn.loc.gov/2020039441

ISBN: 978-0-367-37380-1 (hbk)
ISBN: 978-0-367-37381-8 (pbk)
ISBN: 978-0-429-35341-3 (ebk)

Typeset in Goudy
by Deanta Global Publishing Services, Chennai, India

Contents

List of contributors vii
Foreword xi
Preface xiii

1 Introduction 1
   LIZ MALE AND PENNY NORTON

2 Communications for planning: effective communication through consultation 11
   PENNY NORTON

3 Communications for architecture: telling the stories about our places and cities – communications strategies for architecture and urbanism 39
   DAN GERRELLA

4 Communications for major contractors: embracing twenty-first-century communications challenges 61
   ANDREW GELDARD

5 Communications for specialist subcontractors: demonstrating value in the specialist supply chain 75
   CATHY BARLOW

6 Communications for construction products: promotional practices from specification to installation 91
   LOUISE MORGAN

7 Communications for infrastructure: successful stakeholder engagement on transport and infrastructure projects 107
   JO FIELD

8 Communications for building consultancies: strategies,
   methods and examples   131
   TOM SMITH

9 Communications for construction technology: adopting
   a strategic approach to a rapidly changing sector   147
   PAUL WILKINSON

10 Communications for sustainability: sustainability
   communications in construction and the built environment   158
   LIZ MALE

11 Conclusion: the future of communicating construction   181
   LIZ MALE AND PENNY NORTON

   Glossary   195
   Further reading   212
   Index   213

Contributors

Cathy Barlow first worked for construction specialist PR consultancy Smith Goodfellow in 1985, where she started out as a receptionist with ambitions to be a writer. Two years later, having progressed to become an account executive, she left the business, but continued to work for them on and off as a freelance writer. While bringing up her two children, she took on various part time and voluntary roles, including proof reading the Journal of the Association of Charted Physiotherapists in Obstetrics and Gynaecology, working as a breastfeeding counsellor for the National Childbirth Trust and as a Parent Governor, and becoming the Chair of Stockport's Parent Governor Association and Vice Chair of the North West Regional Parent Governor Association. In 2002 Cathy received a phone call from the founder of Smith Goodfellow asking whether she was still freelancing. The request was to write an article on the issue of hard to heat homes. Some deep research later and she was hooked, coming to the realisation that this work could have an important social impact. Her passion for PR and the built environment has never wavered since. Cathy and her husband Paul became the proud owners and directors of Smith Goodfellow in 2010.

Jo Field is the award-winning founder and managing director of JFG Communications, a boutique consultancy specialising in stakeholder engagement and communications for the transport and infrastructure sectors. Jo is an expert in connecting people and policymakers. She helps companies engage their stakeholders, build advocacy about what they do and inform and influence public policy. Before setting up JFG Communications, Jo built and led Transport for London's award-winning stakeholder engagement team and secured stakeholder support for London's transport infrastructure and the funding to deliver it. Jo has built advocacy for some of the UK's leading transport infrastructure projects, including High Speed 2, Crossrail 1, Crossrail 2, the Tube upgrade and the London 2012 Games. Jo is a Fellow of the Chartered Institute of Public Relations and a trainer on the Institute's stakeholder engagement course.

Andrew Geldard is chief communications officer at Willmott Dixon, a privately owned company that specialises in construction and interior fit-out. Andrew has over 20 years' experience in construction which also includes stints at John Laing and Skanska. He believes that communications, when used effectively, can give companies the 'X' factor for attracting and retaining the best people and also in creating business growth opportunities. He is a passionate believer that companies must have a diverse and inclusive workplace to thrive, and one recent career highlight was seeing Willmott Dixon listed as the UK's number one company in the FT's Diversity Leaders 2020 list, coming third overall out of 700 companies in Europe. Andrew is also a firm believer that companies must be values-led, and helped develop Willmott Dixon's 'purpose beyond profit' ethos, where its people and activities are focused on helping to create a society that works for all. His belief in a company's power to support social mobility saw him devise Willmott Dixon's Foundation in 2011, whose work has improved the life chances of 10,000 people in the last seven years.

Dan Gerrella is a chartered PR professional specialising in property and professional services. Originally qualified as a journalist, he has worked both client-side and in consultancy, supporting organisations in the UK and internationally, mainly throughout EMEA and Asia-Pacific. This has included developing and delivering strategic communications programmes ranging from small, targeted stakeholder engagement through to broader messaging for large, hard to reach audiences, supporting business growth and enhancing corporate reputation. He also has a wealth of crisis communications experience, working with asset owners, property developers and energy companies among others. Dan is associate director at LMC (Liz Male Consulting), a specialist communications and marketing consultancy for the construction and property sector and supports the Chartered Institute of Public Relations through a number of roles. This includes being a member of the CIPR Council, part of the Policy and Campaigns Committee and chair of CAPSIG, an industry group focused on best practice communications within the built environment.

Liz Male MBE is founding director of LMC (Liz Male Consulting), a multi award-winning PR and communications consultancy specialising in construction, property and sustainability in the built environment. Clients range from industry institutes and trade associations, to architects, consultants, contractors and housebuilders, specialist sub-contractors, construction technology suppliers, product manufacturers and the industry's largest events. Over the last ten years, Liz has also undertaken a range of non-executive roles focusing on professional standards, skills and consumer protection within the construction industry. She is chair of Trustees of the National Energy Foundation and a lay member of the Architects Registration Board. She was the independent chair of TrustMark from 2011–2017 and undertook roles on the Construction Leadership Council and the implementation board of the Each Home Counts review to increase skills, competence and consumer protection in the general building and housing retrofit sector. Liz was awarded MBE in the 2015 New Year's Honours for services to construction and consumer protection. Liz has collaborated with Penny Norton on a sister book to this one, *Promoting Property*, also published by Routledge.

Louise Morgan has worked agency-side with construction sector clients for over 15 years. Leading targeted PR programmes which launch disruptive products into the market, drive material specifications and raise the profile of brands throughout the construction supply chain, Louise has been a trusted advisor to many major construction product companies including Lumin, ROCKWOOL, SIG and Trade Fabrication Systems. Louise founded specialist agency Technical Marketing & PR in 2011 with a clear focus on delivering PR and marketing support to the construction and manufacturing sectors. A CIPR and CIM qualified practitioner, Louise holds a BSc in Human Psychology from Aston University.

Penny Norton is a consultation specialist with wide-ranging experience of public relations and public affairs within property, construction and regeneration. Penny has worked with many leading developers, planning consultancies and local authorities. Through ConsultOnline, which she founded in 2012, Penny's work makes pioneering use of social media and web-based communications. She has written extensively on the subject of consultation for property publications. Her first book, *Public Consultation and Community Involvement in Planning: A Twenty-First Century Guide* was published by Routledge in 2017. A Senior Associate of The Consultation Institute, Penny is an active member of tCI's Planning Working Group and jointly set up its Environment Working Group. She has responded to Government consultations, written elearning courses and spoken at property industry events. She is also an active member of the Town and Country Planning Association and has contributed to TCPA best practice guides. Penny has a masters qualification in administration and a postgraduate diploma in PR. She is a Fellow of the Chartered Institute of Public Relations and founded the CIPR's Construction and Property Special Interest Group (CAPSIG).

Tom Smith is an experienced external communications practitioner and has a background of working in infrastructure, development and financial services sectors. Skilled in B2B and B2C media relations, corporate communications and reputation management, he is currently group media relations manager at Mott MacDonald, a £1.5bn global engineering, management and development consultancy. Prior to this, Tom advised some of the UK's largest financial services providers, providing public relations and governmental affairs support and helping to enhance and manage corporate reputations. He has also organised events for the United Nations Conference on Trade and Development.

Paul Wilkinson has worked in UK PR and marketing since 1987, starting in construction professional services before being appointed head of communications at a construction Software-as-a-Service (SaaS) start-up in 2000. Since 2009, he has been an independent technology analyst, PR consultant, freelance journalist, blogger and public speaker. He has written over 1,900 posts on his blog *Extranet Evolution* since 2005 and has freelanced for *AEC Magazine*, *Infrastructure Intelligence* and *Concrete Quarterly*, among others.

An expert on construction collaboration technology platforms, SaaS and related developments in fields including BIM, mobile technologies and social media, Wilkinson is deputy chair of the Institution of Civil Engineers' digital transformation community of practice, executive director and chair of the technology group at the UK BIM Alliance and on the management teams at COMIT and ThinkBIM. A former visiting lecturer in BIM at the University of Westminster, he is also active in Constructing Excellence. He is a Fellow of the CIPR, a past member of its Council and has chaired its policy and campaigns committee and its construction and property group (CAPSIG). He is a member of International Building Press.

Foreword

As this book goes to press, the Public Inquiry into the fire at Grenfell Tower is well into its second phase. With 72 people dead from the worst UK residential fire since the Second World War, there is an understandable desire to apportion responsibility and to ensure that we never see such a grim death toll again.

The evidence before the Inquiry will provide some insight into possible failures in communication that culminated in the disaster of June 2017. When the fire started, did firefighters give residents the right advice? When the cladding was installed on the building, did the specifiers, contractors and building owner understand the risks it might present? When the residents raised their fears with the Council and the Tenant Management Organisation, were they taken seriously? There were many professionals involved in advising, checking, reporting and recommending different courses of action in the years leading up to the fire. Were they competent? Did they understand the implications of their decisions at each stage of the specification, procurement, installation and maintenance of the building's refurbishment? Did they offer the right options to those with the authority to change things? Fundamentally, were the residents and the public generally well-served by the construction industry and its many specialist sectors and advisers?

The Inquiry is, of course, a very large exercise in communicating about construction – witnesses are being heard and statements probed as it tries to work its way towards the truth. Its findings will undoubtedly create a very challenging backdrop that will impact on PR and communications in the built environment for many years to come.

This book is about exploring good practice in communicating construction, and includes chapters covering a range of issues such as planning, architecture, contech (construction technology), major and specialist contractors, infrastructure, building consultancies, building products and sustainability in the built environment.

Each chapter demonstrates how varied each area is within this huge industry, and how much scope there is to demonstrate the power of good PR as a strategic management discipline.

It is impossible to know for now whether better communication could have prevented the Grenfell disaster, but it is at least possible that more open, timely, clearer and better-judged communication could have helped avoid the worst. It is evident that PR professionals have a vital role to play in informing and influencing future change that directly impacts life safety, along with addressing many other industry issues too.

As communication professionals and students of PR, readers of this book have an opportunity in their own working lives to learn, reflect and improve and to ensure that public confidence in construction can be restored. This book provides an excellent starting point to learn how best to approach that work.

The CIPR stands ready to support you too in this respect. For example, the CIPR's own Construction and Property Special Interest Group (CAPSIG) is a forum where those working in the sector can develop their knowledge and their personal network. If you enjoy reading this book, I hope you will come along.

Alastair McCapra
Chief Executive of the Chartered Institute of Public Relations

Preface

The ambition behind *Communicating Construction* was a bold one: to write an introduction to construction PR which would have enough breadth to hint at the huge variety of specialisms within the industry and the scope of PR opportunities, but also enough depth, drawing on extensive experience from across the industry, to provide the reader with practical advice across a wide range of strategy and tactics.

The subtitle *Insight, Experience and Best Practice* reflects the important point that the book is much more than theory. The group of highly experienced practitioners who have contributed provide the context for communications in the sector; they explore the ethical issues; they asses common (and less common) challenges and advise on how they can be mitigated and they cover a wider range of (ever-growing) communications tactics and their practical implementation. We are delighted that case studies, demonstrating all this and more, form a considerable part of the book.

Construction is such a wide-ranging sector. Few areas of our lives are untouched by it, and most issues of the day – from the environment to the economy, the digital revolution to world politics – have a bearing on how it functions and consequently how it is communicated.

Putting aside the devastating impact of the COVID-19 pandemic for a moment (if that is even possible), perhaps some of the most significant challenges facing construction communications today are how it will deal with the impact of the Grenfell Tower tragedy, Brexit, climate change and the ever-increasing skills crisis. It also faces challenges around improving equality, diversity and inclusion and around the adoption of new technologies. And for communicators, there is also the online revolution which is transforming all aspects of PR, providing a plethora of new communications tools and increasing integration with related activities, particularly digital marketing, search engine optimisation, content marketing and advertising.

Unsurprisingly, the majority of PR specialisms – business-to-business, community relations, corporate, crisis management, CSR, investor relations, local and national government lobbying, media stakeholder engagement, public affairs, public consultation, reputation management – are all part of the construction PR professional's toolkit.

The focus on strategy remains paramount. Where a well-researched, strategic approach is undertaken, messages are better formulated, audiences are better defined, work is better coordinated and success is better evaluated.

As fellows and active members of the Chartered Institute of Public Relations, we are strong advocates of the strategic approach to communications as set out by the CIPR and specifically the importance of maintaining a two-way flow of information in any communications programme and the importance of ongoing evaluation and analysis. We are grateful that the CIPR's chief executive, Alastair McCapra, has provided such a powerful call to action in the Foreword, and its 2019 President, Emma Leech, has given us a strong endorsement.

We cannot pretend to have achieved the ultimate ambition of covering every aspect of the complex world of construction or every communications tactic now in existence. But we hope that the rapidly expanding sector will find this book illuminating, informative and inspiring and that it enables the communication of the considerable achievements of a built environment industry we love.

Penny Norton and Liz Male
July 2020

1 Introduction

Liz Male and Penny Norton

Why construction matters

The construction industry is responsible for delivering a huge range of structures that make up the built environment, from Dubai's Burj Khalifa tower, the world's tallest building, to tiny homes and sacred buildings. Anything can make the headlines, be it a huge, award-winning airport terminal to a humble footbridge, from major hospital contracts to self-build projects and domestic repair and maintenance work on our homes.

For each of these areas of construction, there are teams using an array of products and materials and requiring multiple areas of expertise in planning, design, specification, construction, fit-out, management and even demolition.

The word 'construction', therefore, covers a vast array of new build, repair and refurbishment services for infrastructure and buildings, plus the spaces in between, and includes a myriad of specialist activities, consultancies, trades and technologies.

Its range and scale of activity makes this one of the largest sectors in the global economy, with around US$10 trillion being spent on construction-related goods and services annually.[1]

In the UK, the sector contributed £117 billion to the UK economy in 2018, which was 6% of the total economy. In the same year, construction accounted for 6.6% of all UK jobs, with some 2.4 million workers in all, and for 13% of UK businesses, with more than 340,000 registered businesses alone. In addition to these registered firms, the industry has a host of unregistered businesses, which are typically self-employed contractors.[2]

According to the Construction Leadership Council, the industry is six times larger than the automotive industry.

The building cycle is complex, as shown in Figure 1.1 and in a useful video on YouTube from project information provider Glenigan.[3]

Construction projects may be commissioned by clients from the public sector, private sector or partnerships of both. Some clients will commission projects frequently and have in-house expertise and knowledge to bring to a project. Others may be less experienced, only commissioning a single project over decades and so may rely on independent advisors.

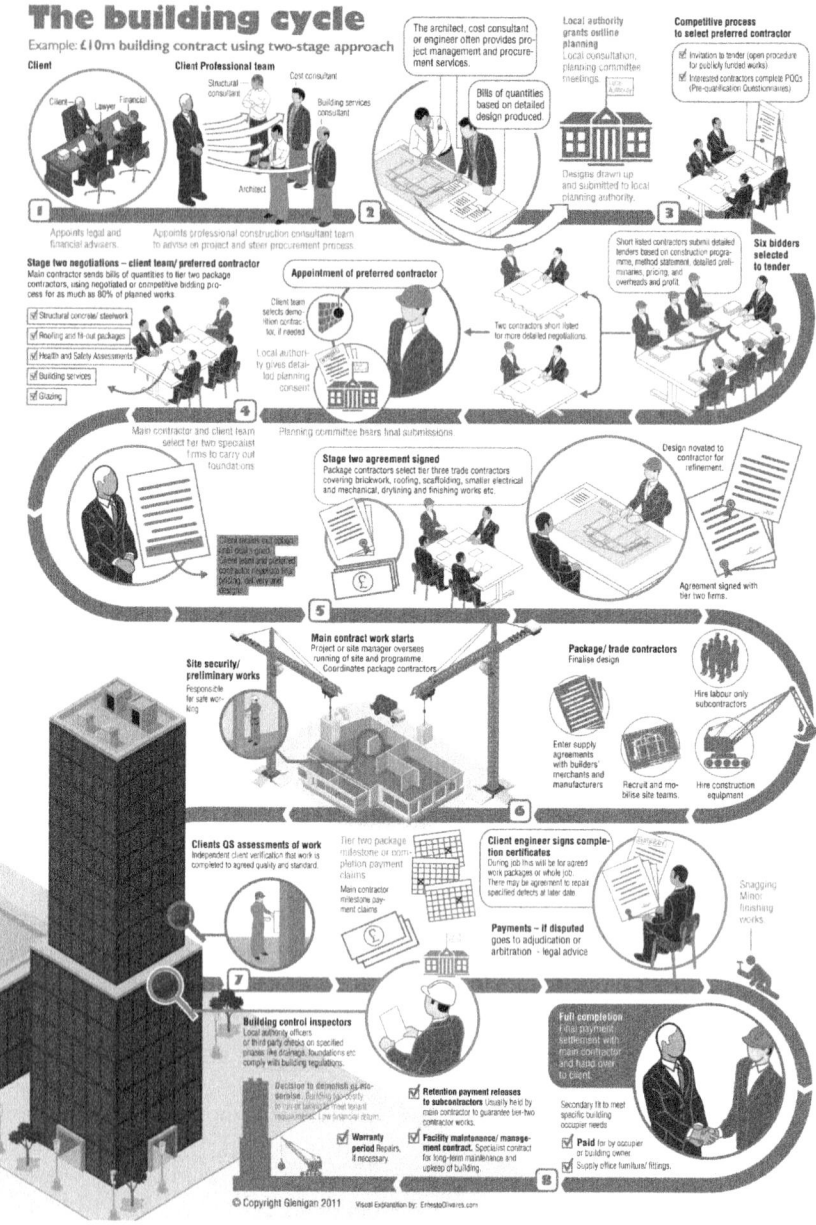

Figure 1.1 An illustration of a typical 'two stage' building contract (Glenigan, 2011). There are multiple other forms of procurement contract in construction. Changes to the building cycle are also being introduced in 2021 as a result of new building safety legislation.

To deliver a project using a traditional contract approach, clients generally draw on the expertise of a team of consultants, advising on such areas as planning, procurement, cost management and design, as well as a main contractor and an extensive line-up of subcontractors to carry out the construction works. The 'Tier 1' main contractor manages the 'Tier 2' subcontractors, who provide specialist expertise in areas such as foundations, structural steelwork, cladding, glazing, building services, roofing, fit-out packages and so on. These professions and skills form their own 'tribes' within construction, having clearly defined areas of responsibility in the project team hierarchy and on site.

An example of the construction process for a new building or structure, and some of the different professions and contracting skills that can be involved, is shown in Figure 1.1.

However, this is by no means definitive, as new buildings and structures can be constructed using a variety of contractual routes, methods and technologies, and their architecture can be highly specific to its context, even when designed using standard components. Existing buildings also present their own challenges in their individual heritage, constraints and client requirements.

The construction industry is, therefore, far from homogeneous in the ways in which it works, its outputs and its workforce. In view of this, it is important for communications professionals to gain an understanding of areas of construction beyond their immediate working field.

This book also begins with chapters on planning and architecture, the two factors that determine and shape construction and take a project from design brief to architectural realisation. There are chapters covering the activities of construction's overarching groups: the main contractors, specialist contractors and consultants. The highly specialised demands of communications for transport and infrastructure projects are considered, while two chapters also focus on areas of major significance for the industry's future and its communications: technology and sustainability.

Facing up to the challenges

The UK construction industry demonstrates its capability and expertise every day in its delivery of extraordinary projects. Some of the most notable achievements of recent years include the London 2012 Olympic Park, Scotland's Queensferry Crossing, the regeneration of Manchester's Victoria station and London's 'super sewer', the Thames Tideway Tunnel.

These high profile examples, and the many less well-known projects, are vitally important to the UK economy and to the successful functioning of its cities and communities.

But the construction industry itself has long had a history of poor productivity. Management consultant McKinsey points out that, when viewed from a global perspective, construction's productivity has trailed other industries for decades.[4] Globally, construction sector labour productivity growth has averaged just 1% over the last two decades. By contrast, manufacturing labour productivity growth has averaged 3.6% over the same period.

In the UK, numerous government-initiated reports have found the industry to be inefficient, slow to modernise and struggling to recruit and retain its workers.

For example, the 2016 report of the Farmer Review of the UK Construction Labour Model,[5] which examined labour force and skills issues, highlighted a string of concerns, including a shortage of skilled workers, a widening gap between skills available and those needed, a poor reputation, inadequate training and a lack of policy and industry oversight.

The industry's strengths, weaknesses, opportunities and threats (SWOT) were summarised in the Government's 'Construction 2025' industrial strategy, published in 2013.[6] See Table 1.1.

The industry is highly fragmented, with around 36% of its workforce being self-employed, which contrasts sharply with the average for the whole economy of 13%.[7] Many jobs in the sector are contracts that involve working on a specific project in its location, so employment can be insecure. This is exacerbated by the industry's boom and bust cycles, which result in a stop-start, inconsistent flow of projects and employment opportunities. All this contributes to construction lacking a highly trained and skilled workforce and becoming overly dependent on migrant labour.

The construction industry's workforce is also ageing, with nearly a third of workers being aged over 50, and is notable for its lack of diversity. Women account for just 15% of the construction workforce[8] and black or minority ethnic workers make up just 4%.[9]

The general public has varying perceptions and experiences of UK construction. Industry bodies the Chartered Institute of Building (CIOB), Royal Institution of Chartered Surveyors (RICS), Royal Institute of British Architects (RIBA) and the Construction Products Association (CPA) commissioned market research company Savanta ComRes to look at public appreciation of and attitudes to infrastructure, construction and building. When asked what best described their personal view towards infrastructure, construction and building, 20% opted for 'very favourable' and 38% for 'fairly favourable'.

However, the public had a different attitude to construction activity, perhaps coloured by their experiences of development in their own communities. More than 70% of overall respondents in the survey agreed with the statement, 'the local public is not consulted enough on major projects', highlighting concerns about and the importance of communications at this stage in the development process.[10]

Communities can be resistant to construction in their neighbourhood, because of the disruption of the process or concerns about the design or use of the end result and consequences, such as increased traffic. Such communities are often proficient in the use of social and online media to make their voice heard at planning and construction stages, and community engagement, via social, online and other routes, can become a focus for communications and PR efforts at this phase.

At the same time, the industry has traditionally been poor at promoting the benefits it delivers in its buildings, as well as in local training, employment and other community initiatives. There are a number of industry initiatives that seek to promote and identify best practice, such as the Considerate Constructors Scheme,[11] which looks to raise standards in site management, efficiency, awareness of environmental issues and neighbourliness.

Table 1.1 A summary of the SWOT analysis in 'Construction 2025' (published by the Government in 2013)

Strengths
- Construction is a key sector to the UK economy, both in terms of value add and employment.
- It also has wider economic significance, enabling businesses to flourish and underpinning other parts of our society's infrastructure including schools, hospitals and homes.
- Construction has a large supply chain, and most of its spend tends to stay within the UK.
- We enjoy world class design skills, particularly in architectural design, civil engineering and sustainable construction.
- Construction has a low entry cost and low capital requirements, allowing small firms to access the market and promoting competition in the sector.

Weaknesses
- Vertical integration in the supply chain is low and there is high reliance on sub-contracting. Lack of integration often leads to fracture between design and construction management and a fracture between the management of construction and its execution, leading to lost opportunities to innovate.
- Investment in R&D, innovation and new processes is low, particularly among sub-contractors, due to uncertain demand for new goods and limited collaboration.
- There is a lack of collaboration and limited knowledge or technology sharing, with learning points from projects often lost when the team breaks up and a project ends.
- In comparison to foreign competitors, UK construction is costly. There are significant opportunities to reduce high costs through greater use of technology, new materials and innovation.

Opportunities
- There are large growth opportunities for UK construction businesses in emerging markets, both in products and high value services.
- There are substantial opportunities for the UK's low carbon construction expertise, both in domestic and foreign markets due to environmental requirements and greater societal demand for greener products.
- Wide implementation of BIM technologies can improve sector productivity and lower costs due to improved information flow and greater collaboration.
- The industry is capable of delivering its product at substantially lower cost through greater efficiency and greater technology and information sharing.

Threats
- SMEs in construction face more difficulties in accessing bank finance than other sectors. Late payment is a problem. Companies are often unaware of support available to them.
- Skill shortages remain a major threat, especially with low investment in training and shortages among trade and professional occupations, inhibiting technology deployment and innovation.
- There are still difficulties in recruiting new talent into construction, due to perceived low image, lack of gender diversity, low pay and job security due to cyclical nature of demand for construction.
- The UK remains a net importer of construction products and materials, and has not yet specialised in construction exports despite its capabilities and relatively higher proportion of construction-related patents compared to competitors.
- Construction remains very fragmented relative to other sectors and countries, which impacts on levels of collaboration, innovation and its ability to access foreign markets.

Government legislation is also extending the local benefits of construction. The Public Services (Social Value) Act,[12] which came into force in England and Wales in 2013, can provide a framework for positive engagement between the construction industry and the communities in which it works. The legislation requires clients commissioning public services to look to secure wider social, economic and environmental benefits. Contractors delivering construction projects for such clients are able to offer a range of benefits, including training or employment opportunities for specific groups within a community or trade with local suppliers of products, materials and services.

Transforming industry

In 2013, the government and the construction industry set out a joint industrial strategy,[13] which is intended to improve the latter's global competitiveness and potential.

This sets out a vision for the construction industry in 2025, which envisages that it will be:

- known for its talented and diverse workforce
- efficient and technologically advanced
- leading the world in low carbon and green construction exports
- driving growth across the economy
- having clear leadership from the Construction Leadership Council[14]

The specific targets set under the strategy are:

- 50% reduction in the overall time taken to deliver new build and refurbishments
- 33% reduction in the initial cost of construction and the cost of built assets over their lifetime
- 50% reduction in greenhouse gas emissions in the built environment
- 50% reduction in the trade gap between exports and imports for construction products and materials

The industrial strategy and the Farmer Review have informed a construction deal between government and industry. This deal is one of a series of partnerships between government and a number of industries, aimed at creating opportunities to boost productivity, employment, innovation and skills. The Construction Sector Deal,[15] which was launched in 2018, sets out measures for government as well as industry in working towards the industry improvement targets set out in the industrial strategy.

The deal highlights three strategic areas for action:

- offsite manufacturing technologies
- digital
- whole life asset performance

Construction has long been associated with the trades and traditional processes carried out on site, such as bricklaying, carpentry and electrical. Increasingly, these processes are moving off site and into the factory, where quality can be more easily controlled, the weather does not impede production and the workforce can be equipped with specialist skills.

These Modern Methods of Construction (MMC) allow work carried out on site to be more akin to an assembly process, with multiple benefits in speed, quality, safety and resource efficiency.

MMC itself has been further categorised to help the industry understand the different options available to them.[16] In particular, the definition framework is designed to help the mortgage finance, insurance and valuation communities in better understanding and supporting the greater use of MMC across residential development.

The definition framework identifies the following seven MMC categories:

- Category 1 – Pre-manufacturing: 3D primary structural systems
- Category 2 – Pre-manufacturing: 2D primary structural systems
- Category 3 – Pre-manufacturing: non-systemised structural components
- Category 4 – Pre-manufacturing: additive manufacturing
- Category 5 – Pre-manufacturing: non-structural assemblies and sub-assemblies
- Category 6 – Traditional building product-led site labour reduction/productivity improvements
- Category 7 – Site process-led labour reduction/productivity improvements

In time, each of these will probably need its own book to explain how they can be communicated and promoted. They include some of the more cutting-edge construction systems in development today that will one day be entirely commonplace in the industry.

As indicated by some of the most exciting Category 7 solutions, advanced technologies are also changing construction. Robotics firms are deploying robots for such tasks as painting road markings and accessing hard-to-reach locations in existing buildings, while drones are commonly used to carry out surveys more speedily and safely. Digital transformation is moving forward with a government drive for greater deployment of building information modelling (BIM), which involves creating a digital representation of a building to inform its design, construction and operation, and even 'digital twins' of entire cities.

Whole-life asset performance involves shifting the focus away from the initial cost of construction to the costs of a building across its life, taking on board the costs of operation, in particular its use of energy.

With 80% of buildings in use in 2050 expected to have been built over earlier decades and centuries, the industry also needs to focus attention on improving the performance of existing buildings, with energy efficiency again being a priority.

The Construction Leadership Council (CLC) was established in 2013 and is steering the industry to deliver on its commitments in areas including training, procurement and innovation.[17] Such moves will help to transform the industry, to the benefit of businesses, their workforces and the overall built environment.

Its current three main areas of focus are:

- Digital – delivering better, more certain outcomes by using BIM-enabled ways of working
- Manufacturing – increasing the proportion of off-site manufacture (MMC) to improve productivity, quality and safety
- Whole-life performance – getting more out of new and existing assets through the use of smart technologies

Co-chaired by the Construction Minister and the chief executive of the Thames Tideway project, the CLC works through multiple work streams (such as supply chain and business models, skills, smart technology, innovation in buildings, exports and trade, the Green Construction Board and communication).

The CLC has also had a pivotal role in helping the industry to deal with the impacts of the COVID-19 pandemic since March 2020, including publishing Site Operating Procedures and the influential Roadmap to Recovery.[18]

The PR and communications opportunities

So who is going to help communicate all this and help shape the reputation and growth of the industry for the future?

Construction is an important market for UK PR and communications services, with approximately 17% of the PR and communications industry's 95,000 people working in property and construction.[19]

PR and communications services enable the construction industry to make the most of its opportunities, charting its progress, achievements and successes, while also supporting its businesses and projects to achieve their business and social objectives.

Many different sub-disciplines of PR are deployed across the different phases, including:

- business-to-business PR (B2B)
- community relations
- consumer PR (B2C)
- corporate PR
- crisis management
- corporate social responsibility (CSR) – more commonly referred to now as environmental, social and governance (ESG)
- education and arts work
- sponsorship
- financial PR
- local government relations
- media relations
- public consultation
- social and online media
- stakeholder engagement

Public relations and communications roles in the sector are varied and operate at many different levels – from large in-house teams with press officers and digital experts, led by PR professionals operating at board level, to specialist PR consultancies working with clients' marketing managers and MDs. At the smaller end of the scale, sometimes PR is handled by a practice manager or it becomes a part-time responsibility for a junior member of staff.

This book, and its partner publication, *Promoting Property: insight, experience and best practice*,[20] explains how these PR disciplines operate in the various stages and areas of construction. It is designed to help all of these people, at every level in the industry, but particularly:

- students of PR wishing to work in construction
- students of construction-related degrees anticipating an element of PR in their work
- recent graduates drawn to a PR career in the built environment
- in-house and consultancy staff looking to increase their understanding of how PR works in key areas of the UK construction industry

As this book will show, clearly there are multiple opportunities for the PR and communications sector to bring its advice and expertise to support the construction industry as it changes its culture, modernises its working practices and improves its image.

There are few sectors more fascinating or with so much scope for growth. Once a PR practitioner gets a feel for the people within the industry, the huge diversity of activities and various ways in which supply chains work and the issues that touch every part of construction, it is almost impossible to feel bored or short of learning opportunities.

We hope that this collection of views about PR across a variety of aspects of construction will provide advice, guidance, inspiration and an opportunity for reflection.

Notes

1 McKinsey Global Institute. Reinventing construction: a route to higher productivity, February 2017: https://www.mckinsey.com/~/media/McKinsey/Industries/Capital%20Projects%20and%20Infrastructure/Our%20Insights/Reinventing%20construction%20through%20a%20productivity%20revolution/MGI-Reinventing-Construction-Executive-summary.ashx [Accessed 4 June 2020].
2 House of Commons Library Briefing Paper Number 01432, Construction industry statistics and policy, 16 December 2019: https://commonslibrary.parliament.uk/research-briefings/sn01432/ [Accessed 7 June 2020].
3 Glenigan YouTube channel: https://youtu.be/zx9ZkFrhPuE [Accessed 16 August 2020].
4 McKinsey Global Institute. Reinventing construction: a route to higher productivity, February 2017: https://www.mckinsey.com/industries/capital-projects-and-infrastructure/our-insights/reinventing-construction-through-a-productivity-revolution [Accessed 8 June 2020].

5 Construction Leadership Council. Farmer, M. The Farmer Review of the UK construction labour model. Modernise or die: time to decide the industry's future, October 2016: https://www.constructionleadershipcouncil.co.uk/wp-content/uploads/2016/10/Farmer-Review.pdf [Accessed 9 June 2020].
6 HM Government, Construction 2025: https://assets.publishing.service.gov.uk/government/uploads/system/uploads/attachment_data/file/210099/bis-13-955-construction-2025-industrial-strategy.pdf [Accessed 16 August 2020].
7 House of Commons Library Briefing Paper Number 01432, 16 December 2019: https://commonslibrary.parliament.uk/research-briefings/sn01432/ [Accessed 5 June 2020].
8 House of Commons Library Briefing Paper Number CBP06838, 4 March 2020. Women and the Economy. https://commonslibrary.parliament.uk/research-briefings/sn06838/ [Accessed 5 June 2020].
9 Building. The face of diversity: construction's hidden problem, 5 April 2019: https://www.building.co.uk/focus/the-face-of-diversity-constructions-hidden-problem/5098826.article [Accessed 5 June 2020].
10 CIOB. The real face of construction 2020: https://policy.ciob.org/wp-content/uploads/2020/02/The-Real-Face-of-Construction-2020.pdf [Accessed 7 June 2020].
11 Considerate Constructors Scheme: https://www.ccscheme.org.uk [Accessed 5 June 2020].
12 UK Government Public Services (Social Value) Act 2012: http://www.legislation.gov.uk/ukpga/2012/3/enacted [Accessed 7 June 2020].
13 HM Government Industrial strategy: government and industry in partnership, Construction 2025, July 2013: https://assets.publishing.service.gov.uk/government/uploads/system/uploads/attachment_data/file/210099/bis-13-955-construction-2025-industrial-strategy.pdf [Accessed 5 June 2020].
14 Construction Leadership Council: https://www.constructionleadershipcouncil.co.uk [Accessed 5 June 2020].
15 HM Government, Industrial strategy, Construction sector deal, 2018: https://assets.publishing.service.gov.uk/government/uploads/system/uploads/attachment_data/file/731871/construction-sector-deal-print-single.pdf [Accessed 7 June 2020].
16 Modern Methods of Construction working group: developing a definition framework: https://www.gov.uk/government/publications/modern-methods-of-construction-working-group-developing-a-definition-framework [Accessed 16 August 2020].
17 Department for Business, Energy & Industrial Strategy, Policy paper. Construction sector deal: one year on, 22 July 2019: https://www.gov.uk/government/publications/construction-sector-deal/construction-sector-deal-one-year-on [Accessed 7 June 2020].
18 Construction Leadership Council, Roadmap to Recovery, June 2020: http://www.constructionleadershipcouncil.co.uk/wp-content/uploads/2020/06/CLC-Roadmap-to-Recovery-01.06.20.pdf [Accessed 16 August 2020].
19 PRCA Census 2019: https://www.prca.org.uk/sites/default/files/PRCA%20PR%20Census%202019%20v9.8.pdf [Accessed 22 August 2020].
20 Oxford: Routledge, 2020. Norton, P and Male, L, Promoting property: insight, experience and best practice.

2 Communications for planning
Effective communication through consultation

Penny Norton

Introduction to planning

People and planning are inseparable; planning exists to create well-functioning spaces for people, to enable social, economic and environmental priorities to shape places and to protect the natural and cultural heritage for future generations. As such, planning is intrinsically linked to individuals' homes, sense of place and local identity – and consultation, as the means by which we communicate with people on planning issues, is imperative in maintaining good public relations.

Because people are intrinsic to planning, the profession involves people at every stage. Yet planning is essentially about delivering change, and change is rarely popular. There is also a need to balance complex social and political concerns and to facilitate mutually beneficial coalitions between stakeholders. Communications in planning, therefore, is challenging.

There are a variety of ways in which people can be involved in the planning process. Community involvement or community engagement are the terms most commonly used to describe ongoing, informal communication between a developer and a community, while consultation refers to the more specific process of involving a community in shaping proposals or seeking feedback on specific proposals.

Consultation – which takes place at several stages throughout the planning process as shown in Table 2.1 – is described by The Consultation Institute as 'The dynamic process of dialogue between individuals or groups, based upon a genuine exchange of views, with the objective of influencing decisions, policies or programmes of action', a definition which could equally be applied to PR.

This chapter will focus specifically on the means by which developers can consult effectively with local residents. It will address how a strategic approach to consultation, culminating in an appropriate selection of tactics, can enable a dynamic dialogue, and how the feedback received can be used to influence decisions.

Consultation in planning is changing rapidly, and therefore, this chapter looks at the impetus for change and its benefits and drawbacks, along with some insights into forward-thinking methods of consultation.

12 Penny Norton

Table 2.1 Stages of consultation in planning

National government	Legislation (Acts of Parliament) Policy (for example, the National Planning Policy Framework) Planning Policy Statement for NSIP[1] applications
Regional government (London and devolved authorities)	Non-spatial strategy plans (such as transport plans)
Local planning authority	Strategic planning (Local Plans, District Plans) Local policies Recommendations to grant planning permission on masterplans and individual properties
Parish councils and neighbourhood forums	Neighbourhood Plans
Developers	Masterplans Planning applications

As a people-orientated process, planning (and consultation specifically) is not immune to the human characteristics of influence, bias, self-interest and political interference, and so the final section of the chapter addresses some of the most frequently experienced problems and the ways in which these can be addressed. This leads to a set of principles by which I would recommend that any public consultation is managed by organisations and their PR professionals.

Developers' requirement to consult

In most stages of the planning process, as described in Table 2.1 above, there is a clear legal requirement to consult; the Planning Act 2008[2] and Neighbourhood Planning Act 2017[3] respectively put in place stringent procedures for Nationally Significant Infrastructure Projects (NSIPs) and Neighbourhood Planning, and the 2004 Planning and Compulsory Purchase Act[4] determines the way in which local authorities should consult, both on strategic planning and in determining individual planning applications.

For developers submitting a planning application to a local authority, however, there is an absence of legislation, and the requirement to consult is instead limited (in most cases) to recommendations within the National Planning Policy Framework (NPPF).[5] The NPPF states that community involvement should be central to developing new housing; sites should be identified and developed where possible 'with the support of their communities'. It requires engagement specifically with local communities in the context of design policies, which should be developed to 'reflect local aspirations', and states that when dealing with proposals, early engagement with local communities and relevant consultees is encouraged, as it has the potential to 'improve the efficiency and effectiveness of the planning application system for all parties'.

Despite the lack of legislative requirements, the NPPF says local authorities should 'encourage any applicants who are not already required to do so by law to engage with the local community'. Furthermore, the document states that,

> Early engagement has significant potential to improve the efficiency and effectiveness of the planning application system for all parties....Good quality pre-application discussion enables better coordination between public and private resources and improved outcomes for the community... Local authorities... cannot require that a developer engages with them before submitting a planning application, but they should encourage take-up of any pre-application services they do offer

and 'where they think this would be beneficial, encourage applicants...to engage with the local community before submitting their applications'.

In such circumstances where consultation is advised through policy rather than required by law, the expected level of developer engagement is set out in the relevant planning authority's Statement of Community Involvement and site-specific expectations are confirmed in pre-application meetings.

Why consult?

So, while the legal requirement for developers to consult remains opaque, the notion that community involvement can benefit planning decisions is unequivocal.

Communication regarding a proposed planning application is usually the first point of contact that a developer has with the local residents likely to be impacted by its proposals. As such, it is a very important first step in public relations – both in terms of reputation management and community relations. Run well, a consultation can establish lasting relationships with a local community that will reap wide-ranging benefits throughout the construction and marketing stages and beyond. But if done half-heartedly, consultation will attract criticisms of tokenism and sham, a significant blow to an otherwise untarnished reputation.

Strategic consultation

As with all elements of public relations, a strategic approach is the basis of a successful consultation.

A common mistake is for a strategy to become a retrospective document; the team launches into consultation tactics (perhaps based on past practice, experience or recommendation), results are collated, and then in a need to create a meaningful consultation report, a 'strategy' is drafted to justify the approach. Worse still, 'consultation' is merely part of a 'plan, announce and defend' approach which puts in place a development proposal and uses communication to quash any negative sentiment. This top-down approach makes scant use of genuine strategic consultation.

The logical sequence of a strategy can, in contrast, result in a well-informed approach, establish a clear direction which can be shared across the development team and provide the means of identifying appropriate dialogue methods and a framework by which consultation is evaluated.

Table 2.2 describes the basic facets of a communications strategy and the corresponding elements of a typical planning consultation.

A strategic approach to consultation requires a symmetrical flow of information between a potential developer and the local community and must prioritise continual engagement, allowing development proposals to evolve in line with feedback and for the process to adapt where necessary. The strategic framework is not a 'to do' list, but a cycle; situational analysis, issues analysis and stakeholder database benefit from ongoing development. Regular monitoring influences the ongoing selection of dialogue methods, and regular evaluation reinvigorates the strategic direction, as shown in Figure 2.1.

Changes in public consultation

In parallel with the broader communication scene, consultation has undergone substantial change during the first part of the twenty-first century.

Increased mobility has led to changes in community structures, with the concept of the 'community of interest' increasing, as residents' attachment to their geographical communities reduces. The housing shortage, new transport links and increased environmental awareness impact on the way in which space is perceived and protected. And we are experiencing a considerable increase in the power of the protest, partly led by international movements such as concern about climate change but also as a result of the political disconnect which followed the 2016 Brexit vote.

Individuals' voices are increasingly heard both through single issue and direct action groups. Communication between organisations and their publics has improved significantly since the advent of corporate social responsibility in the 1990s, and consequently, individuals have greater expectations of dialogue, openness and transparency.

This is partially driven by an increasingly litigious culture; the number of consultations challenged in the courts has increased substantially, resulting in new guidelines and regulations. Community engagement and consultation is now a legal requirement in many sectors beyond property and construction, and while legislation is not compulsory in all aspects of planning, expectations of consultation are increasing.

The online consultation revolution

Perhaps the most significant factor in the changing consultation landscape is technological change, which impacts the amount of information in circulation, the speed with which it travels and the potential for a message to spread. Online consultation has the potential to reach a wide audience, whether geographically or terms of those who might be unable to attend traditional consultation

Table 2.2 The strategic process

Stages of the strategic process	Purpose	Methods
Research Situational analysis[6]	Gain background information to inform the consultation.	Consider the possible influence of a wide range of factors. External issues include policies and political sensitives, attitudes towards the principle of development and the potential for consultation fatigue. Internal factors include resourcing, past experience, local links and the developer's existing reputation.
Issues analysis	Develop understanding of the issues likely to influence feedback on the specific proposals, adding context to the consultation responses and enabling the consultation team to address any misapprehensions.	Utilise town/parish council and local authority information and meeting records, local press coverage and reports/comment from local newspapers, blogs websites and social media. Research imminent and recent planning applications and view responses to consultations.
Political analysis	Understand the political forces which may influence the consultation.	Research community/political/religious and special interest groups, their leadership, membership, policies and influences. Research the political make-up of the council and its members, specifically those with relevant Executive roles, planning committee members, ward councillors and others with significant influence. Check the timing of future elections. Read key documents such as the Statement of Community Involvement, Local Plan and Neighbourhood Plan.
Stakeholder analysis	Understand the local community and the personalities and groups which shape it.	Utilise Office for National Statistics/Acorn consumer classification. Speak to the local authority's community liaison officer to identity both potentially interested groups and those classified as 'hard to reach'. Use stakeholder mapping to determine the communications needs and expectations of specific group and individuals so that they may be targeted appropriately.

(*Continued*)

Table 2.2 Continued

Stages of the strategic process	Purpose	Methods
Scoping Pre-consultation dialogue	Begin a process of constructive dialogue with key stakeholders.	Hold initial meetings with community leaders to develop an understanding of the issues and discuss the intended approach. Identify those who might be engaged as a bridge to the local community.
Strategy Aims and objectives	Develop clear consultation goals to ensure consistency within the development team and communicate a sense of purpose to external audiences.	Jointly determine what the consultation is intended to achieve and identify measurable objectives.
Messages	Ensure consistency in communication	Create an internal document which states the aims and objectives and also addresses the issues identified previously. (A set of Frequently Asked Questions can be useful for setting out issues alongside an agreed response.)
Questions	Identify the questions that will deliver the necessary information.	Draft questions taking into account which aspects of the proposal can be changed in response to local views and the information required. Take into account how responses will be analysed and always test questions.
Target audience	Identify precisely who will be targeted so that resources can be focussed appropriately and monitoring can identify any gaps.	Maintain a database of local residents and interested groups (taking into account GDPR[7] restrictions).
Strategic overview	Clarify the agreed strategy	Ensure that the development team and local planning authority are in agreement about the extent of the consultation and the aspects of the planning application on which local residents can input. Draft a consultation mandate which clarifies: • The organisation running the consultation • The target audience • The aims and objectives • The subject for discussion • The way in which results will be used

(*Continued*)

Table 2.2 Continued

Stages of the strategic process	Purpose	Methods
Selection of dialogue methods	Select dialogue methods	Consider the wide range of dialogue methods available and use the knowledge gained previously to identify those most suitable.
		Take into account accessibility, appeal, a balance of innovation and more established methods, past successes and failures, the time available and the means of analysis.
		Include those which offer information, and those intended to gain both quantitative and qualitative responses.
		Take into account the information requirements of the consultation, specifically whether contact and demographic data is beneficial or whether the consultation will accept anonymous responses.
Resource allocation	Ensure that the proposed consultation is deliverable	Cost the proposed consultation, taking into account human and financial resources.
Create the necessary consultation documents	Ensure that consultees have adequate and appropriate information available to them.	Create informative resources (consultation papers, exhibition boards, website and email text) which provide enough information to allow intelligent consideration.
		Public-facing documents should be clear, transparent and free from jargon.
		Inclusion of a consultation mandate will provide the necessary introduction to the consultation, but the documents should also provide broader contextual information, explain the proposals in detail, present the options available and provide a means for response.
Timetable	Plan for the consultation to be completed in the time available and that adequate time is given to allow for responses, analyse results and publish feedback.	Taking into account holiday periods, create a timetable which details the consultation methods, the timing, roles and responsibilities.
Monitoring	Continually check that the consultation is running as intended.	Ensure that the consultation meets the agreed aims and objectives and is successful in reaching a wide range of local residents. Be prepared to make changes if necessary.

(Continued)

18 *Penny Norton*

Table 2.2 Continued

Stages of the strategic process	Purpose	Methods
Analysis	Gain an understanding from the consultation responses.	Use quantitative and qualitative analysis to inform the development team of the feedback.
Report	Demonstrate the impact of the consultation on the development decision.	Present (or summarise) the analysis and demonstrate how feedback has influenced decision-making. Ensure that all personal data is redacted to comply with GDPR / data protection law. Consider publishing a summary of the report for ease of access.
Respond	Thank respondents and assure them that their responses have been constructive.	Use both direct (one-to-one) and general communication to direct consultees to the consultation report and inform them of the next steps in the development process. Consider a press release to announce the results of the consultation and include the consultation report on the consultation website.
Evaluation	Assess the consultation to address any criticisms and benefit future consultations.	Review success based on the original aims and objectives.

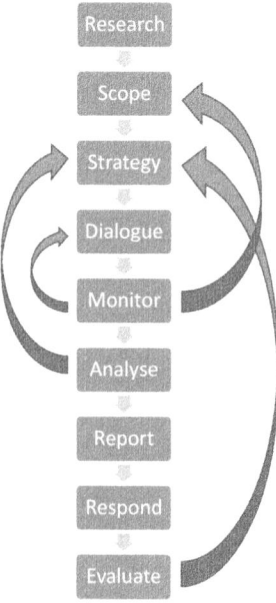

Figure 2.1 The continual approach to strategy development.

events. Online, a level playing field reduces hierarchy; communication can be on a one-to-one, one-to-many and many-to-many basis, and the filter of the traditional media is reduced. Sophisticated communications tactics can allow for the formation of ideas or concepts, a more iterative approach, more responsive dialogue andgreater flexibility. Furthermore, new methodology enables efficient and extensive consultation analysis.

Dialogue methods

Writing about change in consultation in 2016,[8] I identified that online consultation fell into three categories: via social media (primarily Facebook, Twitter and YouTube), using off-the-shelf consultation websites (Citizen Space, Bang the Table) and through bespoke consultation websites (either produced in-house or using an adaptable template website such as ConsultOnline, possibly with additional third-party plugins and widgets embedded).

But in just three years, the selection of dialogue methods available has proliferated to the extent that to define by category would date immediately.

Box 2.1 Online consultation platform: O&H Properties, Marston Valley

Liz Male Consulting (LMC) was appointed by O&H Properties, a major land owner and master developer, to run a public consultation on the masterplan for a mixed-use development and used ConsultOnline to run the online consultation.

Online consultation supported the consultation's objectives to reach a wide geographic and demographic profile quickly and efficiently, to target all age ranges, to convey information through a variety of means, to enable people to communicate at a time and place of their choice and to produce reliable data quickly and efficiently.

Information was available as text, video, images, documentation, email updates and Google Maps. Dialogue took place through questionnaires and the posting of questions online. The service was monitored 24/7, enabling ConsultOnline to become aware of, to understand and to correct misconceptions immediately, and for those taking part to receive a quick response. A user guide provided step-by-step advice in addition to terms and conditions.

The website attracted in excess of 1,200 unique users, with a significant proportion signing up to register for news updates.

The website functionality enabled members of the development team to document off-line discussions with individuals and community groups via a specific portal, enabling ConsultOnline to provide O&H Properties with comprehensive regular analysis comprising both online and offline consultation responses.

20 *Penny Norton*

Anyone may view this website but if you would like to submit your views, ideas and questions through this site, we need you to register online.

Register today to:

- Give us your views
- Take part in polls
- Ask a question to go into the Q&As page
- Put forward ideas for facilities you'd like to see in Marston Valley
- Propose community groups that need our support
- Keep updated on the planning process

Your personal details will be kept safe and in compliance with all relevant data protection regulations.

REGISTER NOW

Figure 2.2 ConsultOnline consultation website (Marston Valley).

Communications for planning 21

1. Which types of home do you feel are needed in the area?
- Apartments
- Family homes
- Retirement homes / elderly accommodation
- Affordable homes
- Self build plots
- Bungalows
- Other

2. If the principle of new growth is confirmed in the Local Plan, are you broadly in favour of O&H's approach of creating a series of villages to deliver the number of new homes needed, rather than a single development of 5,000 homes?
Yes
No

3. Our proposals outline some ideas for waterside living. We welcome any views on these suggestions – please let us know what you think or simply post a picture or a link of waterside homes that you've seen elsewhere or online.

Figure 2.3 ConsultOnline consultation website (Marston Valley).

1. There will be around 300 hectares of informal open space throughout the scheme. This is likely to comprise all or some of the following. Please let us know which of the following you value the most by ranking from 1 (highest priority) to 9 (lowest priority):

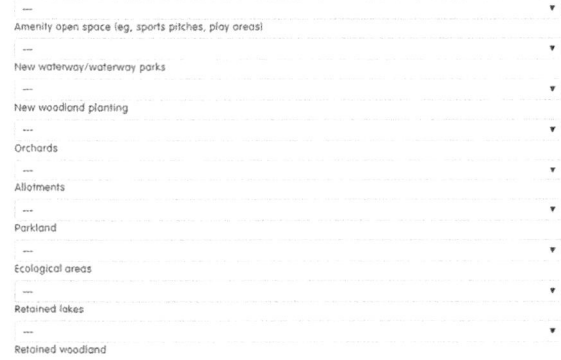

2. Our proposals are shaped by the natural features of the Marston Vale. We plan to create substantial amounts of green open space (eg, woodlands and country parks). Do you support the provision of this facility?

Figure 2.4 ConsultOnline consultation website (Marston Valley).

22 *Penny Norton*

Consultation websites should no longer be seen as single dialogue methods but as increasingly flexible platforms for a range of tactics including polls, forums, infographics, videos, blogs, vlogs, blog posts and podcasts which offer immediate and very effective means of analysis.

In addition to websites, consultation apps are now increasingly used in consultation.

Box 2.2 Consultation app: Built-ID

Built-ID is a web-based platform which encourages collaboration data-sharing at various stages throughout the property lifecycle. It aims to transform how people interact with the built environment and is primarily used by subscribers to discover who is working on a project and to gain contact data.

Give-My-View specifically connects development teams and local communities through a visually appealing and highly accessible website app. It enables the development team to guide and educate the community as the project evolves. An interactive timeline manages expectations while a newsfeed enables them to counteract the spread of misinformation.

Other features include polls, questionnaires and 'Quick facts' which provide information in relation to specific questions. Geofencing (the use of GPS technology to create a virtual geographic boundary) provides the ability to restrict comments to a specific geographical area. To encourage use among younger audiences and simultaneously benefit relationships between the developer and the community, the app provides the opportunity to earn points for engaging and sharing. Points translate into money for a selection of local charities chosen by the local council or developer. Community members are able to browse influenced decisions from previous phases which assists in building trust between the developer and the community.

 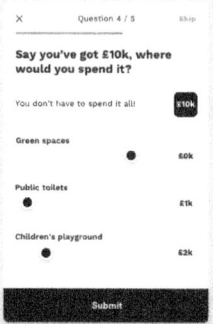 

Figure 2.5 Built-ID consultation app.

Communications for planning 23

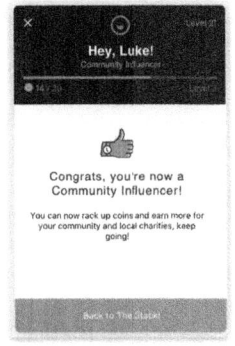

Figure 2.6 Built-ID consultation app.

Online community mapping websites, such as those developed by Commonplace and Bang the Table, enable participants to engage by dropping pins onto a map with comments attached. 3D modelling has enabled developers to create digital models of the proposed development or broader area, and local residents are able to view 3D models from specific locations to understand the impact of change, taking into account density, sunlight paths and building heights. Models can then be adapted to reflect various alternatives that the consultation presents.

The logical extension of this is for augmented reality to project proposed schemes onto actual landscapes. Along with a vision of how the new scheme would fit with the existing infrastructure, augmented reality would enable the user to access additional information data-tagged onto the projected image. Compared to completing a paper questionnaire, technological developments such as these carry significant appeal.

The principles of gaming are also being incorporated into consultation, with Minecraft being used to create and modify communities and with the principles of such games being incorporated into consultation websites.

Box 2.3 Interactive mapping and reporting tool: CHLOE

CHLOE is an online mapping and reporting tool developed by David Lock Associates to enhance engagement and to improve information capture during community and stakeholder workshops.

In small groups, participants are encouraged to create conceptual masterplans for potential growth locations by populating an empty grid, tile-by-tile, with a selection of land uses. An aerial photo overlaid with local constraints helps to inform the group's discussion on where development can and cannot be placed. As development tiles are added to the grid,

24 Penny Norton

CHLOE feeds back live alerts to guide the group through the design process allowing them to make informed decisions about housing numbers, education and open space provision and employment floorspace and/or job delivery to ensure that the development in question is sustainable and that the area's needs are met.

Adjustable parameters, including residential densities, are combined with open space standards and demographic data (specific to each local authority) to highlight the impact that housing has on associated land use composition. Different scenarios, based on changes to the parameters, can be tested, recorded and compared.

CHLOE provides a digital platform for local stakeholders to discuss and consider development that is appropriate, proportionate and represents the existing community's needs.

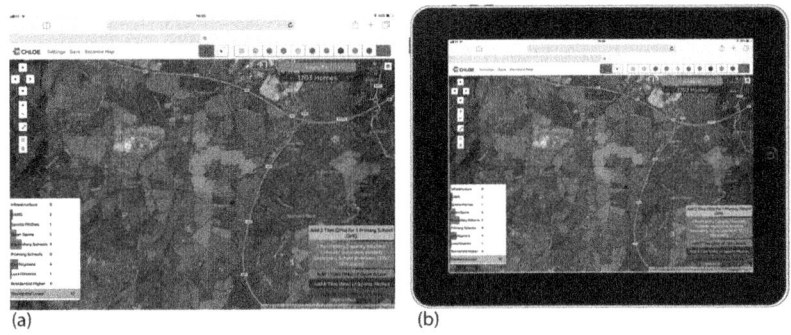

Figure 2.7 CHLOE online mapping and reporting tool.

Box 2.4 CD modelling: VU.CITY

VU.CITY, the product of leading proptech agency Wagstaffs and GIA Surveyors, specialists in rights of light and daylight and sunlight, is the first ever complete fully interactive 3D digital model of London, along with the city centres of several UK and international cities. Through VU.CITY, consultees can visualise proposed developments within the existing context of the city. VU.CITY can demonstrate transport data, overlay sightlines and visualise wind modelling, pollution and sunlight paths. Dropping down to street level further helps consultees understand the proposed new developments in context. VU.CITY also brings in the London View Management Framework in relation to protected views and in relation to existing and consented developments.

Communications for planning 25

British Land's Canada Water development is one example of how VU.CITY and other interactive 3D digital tools have supported the planning and consultation process. At Canada Water, VU.CITY was able to show the architects' proposed masterplan as a 3D interactive model in the wider context of central London, allowing consultees to navigate the platform and understand the proposals in detail. By showing masterplan options, proposed uses, transport links and data overlays, VU.CITY was able to demonstrate in compelling detail, how the proposals would affect the local community.

Figure 2.8 VU.CITY online model.

Figure 2.9 VU.CITY online model.

Online consultation is evolving rapidly, and over the next decade we expect to see increasing use of large format touchables and touchscreens which use geospatial data. Walk-throughs of 3D models and virtual reality theatres will enable a shared experience of a digital representation in a planning workshop. Online 'story maps' which link text and images to locations have the potential to provide information about a proposal either though a desktop app or as part of a walking tour, supported by data on current and proposed land uses. Increasingly photorealistic, high resolution representation of proposals will have the ability to depict alternative scenarios almost equivalent to reality.

Reducing risk in online consultation

Innovation inevitably presents risk. The challenge for the PR professional is to mitigate risk and to capitalise on the opportunities that online consultation presents. Some of the potential risks and concerns in relation to online consultation are examined below.

Initiating, running, moderating and evaluating an online consultation requires skills which may be lacking or in short supply within the development team:

- Be realistic when deciding on consultation tactics; do not take on more than is feasible.
- If using third party software, ensure that the package either includes comprehensive training and support or is run by the supplier.

Consultation websites attract responses from those who clearly have not studied the consultation materials and fully understood the issues:

- Consider increasing the various ways in which information is communicated (for example, including video as an alternative to, or in addition to, text).
- Introduce feedback mechanisms at key points in the text/video to increase the likelihood of consultees responding to a question in context.

Social media makes responses less attributable, and online consultation can be hijacked by organisations which do not form part of the target audience (such as pressure groups):

- When formulating a consultation strategy, determine whether anonymous contributions are acceptable and communicate this information via the consultation mandate.
- Do not attempt to run a consultation on social media; simply use social media to signpost to the consultation website.
- A consultation website can include a requirement for users to register using their postcode, and the electoral register can be used to check the veracity of identities given.

Communications for planning 27

- Alternatively, geofencing can be used to allow only responses from a defined geographic area.

The consultation may be hijacked by trolls:

- Distinguish between anti-social behaviour and a negative response to the consultation. Disappointing though it may be, the latter should not be dismissed. If the protocol is set out in in the consultation mandate or website user guide, posts can be removed with immediate effect, IP addresses banned and usernames invalidated.
- Software used to identify bad language and spam and is advised, in conjunction with monitoring.

A focus on online consultation potentially ignores those who are not internet users:

- Always provide off-line dialogue methods to ensure that non-internet users have equal opportunity to respond.

The organisation is nervous about collecting and storing personal data following GDPR legislation:

- Data protection legislation can result in organisations being fined if data is not held securely; check current guidance.[9]

The development team is concerned about hacking, phishing and spam:

- An unprotected website can leave itself open to abuse, and if user details are being collated via an online database, the legal and reputational impact can be considerable. All websites should include EV (Extended Validation) SSL (Secure Sockets Layer) certificates to ensure that communications, specifically user names and passwords, are encrypted and can only be accessed by the website owner. If you are using a website provided by an external company, choose a provider who is ISO 27001 accredited.

Website accessibility issues can be complex and require additional budget:

- Keep design simple, with minimal and intuitive navigation steps.
- Consider physical accessibility in allowing functionality via either the keyboard or mouse, and ensure that buttons and selectable areas are of sufficient size.
- There are many steps that can be taken to make a website accessible to those with poor vision. These include enabling interactivity with screen readers such as JAWS, NVDA, VoiceOver for OS X, Window Eyes and Supernova, basic operating system screen magnifiers such as ZoomText and MAGic and

ensuring that the website is compatible with speech recognition software such as Dragon Naturally Speaking. Also consider font size, contrasting colours and the use of colour palettes which might be problematic for those with common forms of colour-blindness.
- Provide a text transcript of audio or visual files for those with hearing impairments.
- Provide time for content to be read and understood and avoid having user interface controls that fade out or disappear after a certain amount of time.
- Consider providing translated text or using a translation service.

It is important that the development and consultation teams fully understand the limitations of online consultation, particularly social media. Understandably, tactics as new and as powerful as online consultation can cause concern. But by far the greatest risk in online consultation is not connected to the online platform itself but the absence of it; failure to provide a means by which local residents can discuss a proposed development online can result in the developer being unaware of other online discussions, which can then gather momentum and perhaps only come to light when it is too late to address serious concerns or misapprehensions.

Co-production in place of conflict

A decade ago, consultation in planning was dominated by the dreaded public meeting. In the worst-case scenario, a development team would prepare a masterplan and, bearing down on them from a podium, tell a large group of local residents what was intended, ideally for long enough to minimise comment. But disgruntled locals were invariably present, quick to jump to their feet to oppose the scheme and rally their neighbours into an angry frenzy. Generally, residents would attend only if they objected – what was the point in venturing out to a draughty village hall if you were broadly in support of the proposals? And the local media would attend in anticipation of a dramatic evening, which often resulted in a harmful news story.

Not all public meetings took this form of course; many were constructive, and they continue today in some circumstances. But generally, development teams now seek more engaging and constructive means of consultation.

Co-production, which in planning is usually referred to as participatory planning, community planning, community visioning or collaborative planning, is gaining increasing prominence, its engaging approach providing a welcome alternative to the adversarial public meeting.

Co-production is a significant departure from the old-style public meeting, necessitating the relinquishing of power to local residents, recognising their experience and, essentially, regarding the local community as 'us' rather than 'them'.

Box 2.5 The New Economic Foundation principles for co-production[10]

- Assets: recognising people as assets
- Capacity: building on people's capabilities
- Mutuality: developing two-way reciprocal relationships
- Networks: encouraging peer support networks
- Blurred roles: blurring boundaries between delivering and receiving services
- Catalysts: facilitating not 'delivering to'

The process of co-production varies, but typically, it involves pre-engagement research and dialogue, a community planning day in which groups of residents, assisted but not directed by professionals, create visions and solutions which they then feed back to the larger group and development of a masterplan by professionals following local insight, followed by an exhibition at which the masterplan is formally consulted upon. Citizens' juries and appreciative inquiries are other co-production tactics that have been used successfully in planning.

The benefits of this approach are substantial. Early engagement can create a sense of ownership among the community, build trust with the development team and result in positive sentiment towards change. Participatory planning, because of its variety of tactics and emphasis on facilitation, can involve a range of local voices including those who would not choose to comment otherwise. The process frequently accelerates the masterplanning process, partly because it involves not only local residents, but also politicians and planners.

Box 2.6 Planning for Real®: Pershore Town Plan

Planning for Real is a nationally recognised community planning tool which enables local residents (the inside experts) to be involved in planning for and within their own communities. Used for a range of purposes including strategic planning, masterplanning and the regeneration scheme of an existing neighbourhood, Planning for Real is useful in incorporating social/people aspects into consideration about the physical place.

It is an eye-catching, hands-on process which, by using a 3D model as a focus, enables local people to put forward suggestions to show how an area can be improved, or to highlight specific problems. This is done using pre-written pictorial, colour coded themed suggestion cards or blank cards for people's own ideas.

At follow up workshops, residents, working in small groups supported by outside experts (planning consultants and other members of the project team, also, in some cases, those representing the local authority, health and wellbeing organisations, the Police and youth workers), look at the suggestions that have come out of the engagement events. These are then prioritised and options narrowed down, so that a clear picture of what needs to be done and what is achievable emerges and forms an Action Plan – the planning proposals (draft strategic plan or masterplan).

Most people coming fresh to Planning for Real, whether they are residents or professionals, expect the strongest personalities to dominate, and that different factions will form and fight to the last. But throughout the process, compromise and consensus become easier, not least because everyone's line of vision converges on the subject matter allowing for practical and non-threatening ways of communicating and participating.

The Planning for Real Unit recently completed work on a project to develop a town plan for Pershore in Worcestershire. The Unit provided support to a steering group, which was set up by Pershore Town Council and made up of local resident volunteers with a minority representation from the Council.

The Steering Group worked tirelessly over a 12-month period creating an overall vision and theme visions, setting priorities and developing an Action Plan. The use of a large plan of Pershore, mounted onto polystyrene and colour coded flags at the consultation events was an approach welcomed by the 1,000 residents, of all ages, who participated.

The wealth of information gathered enabled the theme group leaders, their group members and stakeholders to understand needs, concerns, and opportunities and provided a firm base from which to develop, through follow up meetings and workshops, their individual Theme Action Plans. The town plan was produced and has been adopted by the Town Council.

Figure 2.10 Pershore Town Plan consultation event.

Co-production is only effective if a considerable amount of time is invested at an early stage, providing an opportunity for the community to be involved in developing a vision, and it requires considerable faith and an enlightened attitude on the part of the development team. But of those who have taken the plunge, the vast majority would take every opportunity to do so again. In fact, this form of consultation is increasing in popularity in every sector, from Neighbourhood Planning to large-scale mixed-use schemes.

Developments in monitoring, analysis and evaluation

Monitoring, analysis and evaluation are important elements of strategic consultation. Monitoring occurs throughout. Analysis, although it can be ongoing, takes place (or is complete) at the end of the process, and this is followed by evaluation.

It goes without saying that electronic communication has enabled a more scientific approach to each. I recently set up an online consultation platform (see Box 2.1) in which all data collected, from comments in meetings to online polls, was collated, enabling me to present the client with an up-to-date consultation report as often as required, at the touch of a button. It is hard to recall how 20 years previously, the developer would have had little knowledge of resident sentiment until the end of the consultation. Today, issues management is a key strategic element of any consultation, and we have created the tools to easily identify emerging themes, possible misapprehensions and potential ambassadors.

With the transition to participatory planning, consultation data has moved from being predominately quantitative to predominately qualitative. Qualitative data – observations and comments, usually expressed in words rather than in numbers – both provides a context for quantitative data and enables the consulting body to get to the heart of an issue. And again, recent technological developments, specifically in coding and mention analysis, have brought about a more effective means of measurement.

An increasingly litigious scene

So far this century, we have experienced a huge increase in the number of planning consultations ending up in the courts.

Due to the absence of a requirement in many cases, consultation in planning is not governed by strict rules and regulations, but the legal aspect of consultation is important because simply using the term 'consultation' creates expectations which can be challenged in the courts. Training in consultation law is strongly advised for anyone carrying out a planning consultation.[11]

There are several 'layers' of law which affect public consultation:

- Statutory – e.g., The Town and Country Planning Act 1990 Chapter 8; Regulation 10 of the Town and Country Planning Order 1995, the Localism Act 2011, The NPPF, the Planning Act 2008

- The Freedom of Information Act – some of which has been trumped by 2018 GDPR legislation[12]
- Equality – Equal Opportunities Act 2010
- Environmental – the requirement to carry out Environmental Impact Assessments which is based on the 'Three Pillars' of the The Aarhus Convention[13] (Access to information; Public participation in decision-making; Access to justice)
- The Doctrine of Legitimate Expectation – where courts recognise consultees' right to expect a thorough consultation
- Common law – which rests on an increasingly substantial amount of case law

Most case law on regulatory consultation is viewed in the light of the Gunning Principles[14] which set out the legal expectations of what is appropriate consultation and, importantly, provide an extremely helpful means of ensuring that a consultation is sound.

Box 2.7 The Gunning Principles

Gunning 1: when proposals are still at a formative stage

Consultations have been found to be at fault on this basis if a decision has already been made; if the critical question is avoided; if consultees are not consulted on all options; if a single 'over-engineered' option is the only option; if the options are portrayed inaccurately.

Gunning 2: sufficient information to give 'intelligent consideration'

Case law includes promises for an extensive consultation being broken; a lack of transparency in options development; failure of the consulting body to give adequate reasons for decisions made; unclear information; failure to ask the right questions; failure to provide adequate information; proposals not adequately communicated.

Gunning 3: adequate time for consideration and response

Failure at this hurdle has been the result of the consultation process not being visible or effectively publicised; inadequate time being allowed for responses; inappropriate phasing.

Gunning 4: must be conscientiously taken into account

Consultations have been taken to court because of inappropriate weighting of consultation responses; the withdrawal of options before they have been consciously considered; failure to summarise responses adequately; unfair reporting of consultation outcome; failure to consult 'out of area' consultees; failure to re-consult if situations/options change.

> **Box 2.8 The 'three pillars' (Articles 4–9) of the Aarhus Convention**
>
> The Aarhus Convention stipulates three public rights which have become an important benchmark in consultation, specifically in relation to dialogue between the public and public authorities:
>
> 1. Access to information
> 2. Public participation in decision-making
> 3. Access to justice

> **Box 2.9 The Doctrine of Legitimate Expectation**
>
> The Doctrine of Legitimate Expectation originated in the UK and has since become incorporated in the other common law jurisdictions in relation to the practice of public bodies.
>
> A procedural legitimate expectation exists when an organisation commits to following a certain procedure – such as consulting – prior to making a decision. If the expectation to consult is created but not delivered upon, the organisation may lose a Judicial Review on the basis of failing to comply with the Doctrine of Legitimate Expectation.

For the developer not legally required to carry out a consultation there is a simple solution to entering this legal minefield: consider whether contact with local residents need constitute 'consultation', or can instead be labelled 'community involvement' or 'community engagement' – and in doing so, steer clear of the legal and reputational ramifications that can occur should a consultation fail to meet such exacting standards.

Consultation challenges

Taking into account the potential for a consultation to significantly affect a developer's reputation, what might appear to be a lengthy process (even before the consultation begins) and the extent of change impacting on consultation, it is perhaps not surprising that many developers shy away from carrying out a thorough consultation.

However, an understanding of the potential challenges from the outset (which should have been enlightened by thorough issues analysis) will enable the consultation team to address them. Problems with consultation fall into just a handful of categories as shown in Box 2.10.

Box 2.10 Consultation challenges and their causes

Access

- Failure to engage with a wider audience, specifically the 'hard to reach', and to gain responses from the 'silent majority'.
- Apathy and consultation fatigue.

Clarity

- A lack of clarity about the aims of participation leading to disaffection.

Communication

- Failure to explain the situation and its limitations effectively.

Creativity

- A lack of creativity resulting in a lack of motivation.

Disappointing results

- Negative responses, perhaps as a result of campaigning by pressure groups, and negative media involvement.

Failure to respond

- A failure to respond to or act on the outcomes of participation.

Inadequate promotion

- Lack of awareness of opportunities to participate.

Information

- Provision of too much or too little information or failure to simplify complex information.

Managing expectations

- Disappointment in the consultation by those being consulted.

Political interference

- Unwelcome involvement of those with a political agenda beyond the scope of the consultation.

Resistance within the development team

- An internal culture which is inclined to limit consultation, lacks trust in the process, provides too little information too late and fails to listen to feedback.

Resources

- Lack of dedicated resources (people, funding, technology).

Time

- Unreasonable timing, causing a consultation to be rushed, ill-thought-through or otherwise compromised.

Immediately apparent from Box 2.10 is that at least half of the problems likely to arise are in the domain of the PR and communications team; issues relating to access, clarity, communication, creativity, failure to respond, inadequate promotion, information, resistance, resources and time are common issues with communications generally and can each be addressed before the consultation starts.

Advice on how to combat the external issues is provided in greater depth in my earlier book.[15] In writing about addressing the challenges, it became immediately clear to me that almost all problems can be resolved by following the strategic process:

- Situational and issues analysis and pre-consultation dialogue enables the PR professional to identify many of the potential problems that may occur, understand and manage expectations and determine the most appropriate tactics to use.
- Stakeholder analysis will identify the range of local audiences to be involved, from political and community leaders to those regarded as 'hard to reach', and develop an understanding of how best to involve them.
- The aims and objectives, as communicated through the consultation mandate, will help address any criticisms of the consultation in terms of its extent, audiences and use of the results.
- Consistent messaging in the form of a Frequently Asked Questions document will ensure that the whole development team is able to address difficult questions in a public setting and agreement with the local authority over the strategic overview will provide the basis for a good relationship between the development team and local planning authority.
- Resource allocation will prevent issues such as capacity to respond, and monitoring will help identify any problems as they occur so that they can be acted upon quickly.

Finally, monitoring, analysis and evaluation all play an important role in explaining the reasons for consultation results.

Conclusion

The answer to a good consultation is undoubtedly a strategic and principled consultation.

Box 2.11 10 principles of effective consultation

Accessible: provides easy access to all stakeholder groups – access may be cultural, physical, intellectual, technological.

Accountable: assumes responsibility for the impact of the development team's approach to consultation and interpretation of the results.

Engaging: uses appropriate and creative dialogue methods to involve and inspire.

Informative: provides adequate, unbiased information to enable good decision-making, ensure that complex information is explained clearly, and consider adapting consultation materials for different audiences.

Manages expectations: uses pre-consultation dialogue and research and monitoring to understand attitudes and expectations and responds accordingly.

Responsive: responds to all communication quickly and positively; allows development proposals and the process to evolve in line with feedback; ensures that all consultees receive the consultation report.

Strategic: follows the strategic approach to ensure that the consultation is well researched, based on firm objectives and structured and designed to produce meaningful analysis and evaluation. The strategy should be well understood within the development team and communicated to wider audiences in the form of a Consultation Mandate.

Transparent: from setting realistic objectives, communicating the purpose of the consultation, drawing up agendas for discussion and imparting information, to analysing results and providing feedback, openness is paramount. The consultation report provides a clear audit trail of analysis and recommendations so that the impact of consultation upon subsequent decisions is identified. Where feedback which has not been used to inform the final decision, demonstrate the rationale for doing so.

> **Two-way:** aims to achieve a symmetrical flow of information between the development team and the local community, as opposed to bombarding the community with information and paying little attention to responses.
> **Timely:** allows ample time to develop the early stages of the strategy, to engage fully and to provide adequate time for responses. The timescale is set out in a document which can be accessed by all.

Looking ahead, is consultation likely to see such significant change in the next twenty years as it has done in since the Millennium?

Political will is likely to dictate future change. At the time of writing, a Conservative government is responding to both the lack of housing and change on the high street by increasingly introducing permitted development rights. The need not to apply for planning permission in such cases removes the need for local consultation. Simultaneously, the former Labour planning minister, Nick Raynsford, was appointed by the TCPA to review the planning system and the recommendations within his report[16] are for significantly more consultation and community involvement in planning. The future requirement to consult appears to depend on what is currently a very precarious political balance.

Politics aside, and regardless of a legal requirement to consult, consultations are increasingly scrutinised though the courts, necessitating a good awareness of consultation law among all communications teams.

As a result of these trends, and thanks to some excellent training and guidance provided by the Consultation Institute, consultation is becoming increasingly professional. Quality assurance, consultation industry standards of practice, professional accreditations and CPD have contributed to this. In future, I would hope to see the creation of a 'good' consultation kitemark for the industry, increased training for planning consultants on consultation and a formalised means of best practice across the industry, specifically on subjects such as online consultation, evaluation and analysis and the use of co-production. This, together with adherence to the strategic process, will help address the challenges that we currently face.

PR professionals must accept that consultation doesn't necessarily deliver a 'yes' vote in support of a development proposal. But communication with local residents has significant PR advantages – enabling the developer to forge links with local residents and the wider community throughout the lifetime of the project, providing a platform to promote a variety of positive messages – from the use of local suppliers to corporate news stories – and significantly benefiting the developer's reputation. And, since better consultation results in better decisions, the quality of communities, both physically and socially, benefits from effective communication.

Notes

1. A Nationally Significant Infrastructure Project is an infrastructure development above a specific threshold which is identified as being of national importance. Typically this will include railways, highways, water, waste, harbours, power generating stations, wind farms and electricity transmission lines. Such projects typically span several local authority areas, are complex, frequently controversial and often involve compulsory purchase. In response to these limitations – and importantly the fact that these projects are of national significance – the planning system allows for consultation on the principle of the development to occur at a national level, while the local consultation is centred on design, community benefits and mitigating negative impacts to the neighbourhood.
2. Planning Act 2008 https://www.legislation.gov.uk/ukpga/2008/29/pdfs/ukpga_2008 0029_en.pdf [Accessed 12 November 2019].
3. Neighbourhood Planning Act 2017 http://www.legislation.gov.uk/ukpga/2017/20/pdfs /ukpga_20170020_en.pdf [Accessed 12 November 2019].
4. Planning and Compulsory Purchase Act 2004 https://www.legislation.gov.uk/ukpga /2004/5/pdfs/ukpga_20040005_en.pdf [Accessed 12 November 2019].
5. National Planning Policy Framework, pub Ministry of Housing, Communities & Local Government, 2018. https://assets.publishing.service.gov.uk/government/uploads/syste m/uploads/attachment_data/file/733637/National_Planning_Policy_Framework_we b_accessible_version.pdf [Accessed 12 November 2019].
6. Typically PESTEL (political, economic, social, technological, environmental and legal) and SWOT (strengths, weaknesses, opportunities, threats) analyses.
7. The CIPR provides a selection of useful resources for communications professionals on GDPR: https://www.cipr.co.uk/gdpr [Accessed 12 November 2019].
8. Norton, P and Hughes, M *Public Consultation and Community Involvement in Planning* (2018). Abingdon: Routledge.
9. The Government's guide to the General Data Protection Regulation is at https://ww w.gov.uk/government/publications/guide-to-the-general-data-protection-regulation. [Accessed 12 November 2019].
10. https://neweconomics.org [Accessed 12 November 2019].
11. The Consultation Institute runs some excellent training in consultation law which is a must for any consultation professional: https://www.consultationinstitute.org/trainin g-events [Accessed 12 November 2019].
12. The Government's guide to the General Data Protection Regulation is at https://ww w.gov.uk/government/publications/guide-to-the-general-data-protection-regulation. [Accessed 12 November 2019].
13. 1998 United Nations' Economic Commission http://ec.europa.eu/environment/aar hus/index.htm. [Accessed 12 November 2019].
14. The Gunning Principles were put forward by Stephen Sedley QC and accepted by the court in the case of R v Brent London Borough Council, ex parte Gunning (1985) 84 LGR 168 in which Hodgson J quashed Brent's decision to close two schools because the manner of its consultation had been unlawful: 'Mr Sedley submits that these basic requirements are essential if the consultation process is to have a sensible content. First, that consultation must be at a time when proposals are still at a formative stage. Second, that the proposer must give sufficient reasons for any proposal to permit of intelligent consideration and response. Third… that adequate time must be given for consideration and response and, finally, fourth, that the product of consultation must be conscientiously taken into account in finalising any statutory proposals'.
15. Norton, P and Hughes, M *Public Consultation and Community Involvement in Planning* (2018). Abingdon: Routledge. Chapter 18: Reducing Risk in Consultation.
16. Planning 2020: Raynsford Review of Planning in England Final Report (2018). London: TCPA.

3 Communications for architecture

Telling the stories about our places and cities – communications strategies for architecture and urbanism

Dan Gerrella

Introduction to architecture

Architects design buildings and infrastructure. In the UK, they follow the RIBA Plan of Works, an eight-stage process from the strategic definition of a project through to post-occupancy stage (see Figure 3.1).[1]

The common career path involves undergraduate and postgraduate qualifications, two separate placements within practices and a written exam and interview at the final stage (Parts 1–3).[2]

All practising architects in the UK must be registered with the Architects Registration Board (see Box 3.1). They may also choose to become members of the Royal Institute of British Architects (RIBA) or equivalent bodies in Scotland, Wales and Northern Ireland. All have a Code of Conduct.

For those working in communications within this sector, it is important to understand this route and the standards that architects must work to. This is a serious profession, and the people within it are highly qualified. It is the PR professional's job to translate the architect's approach to design into something that resonates with audiences. Clarity of message is more important than technical detail.

There are three main types of architecture practice:

- **Commercial:** the majority of practices. They work across multiple sectors and will deliver 'volume' work as well as one-offs.
- **Signature:** design icons and special buildings. High design standards. They are often hired on the strength of the founding partner's name and reputation.
- **Specialist:** portfolio is predominately made up of one type of building (for example, stadia), or an area of expertise (for example, heritage refurbishment).

As well as architecture, some practices provide services including masterplanning (also known as urban design, this is the design of everything from mixed-use districts through to major cities), interior design, landscaping and branding, graphics and wayfinding. The latter includes elements such as visual identities and on-site signage.

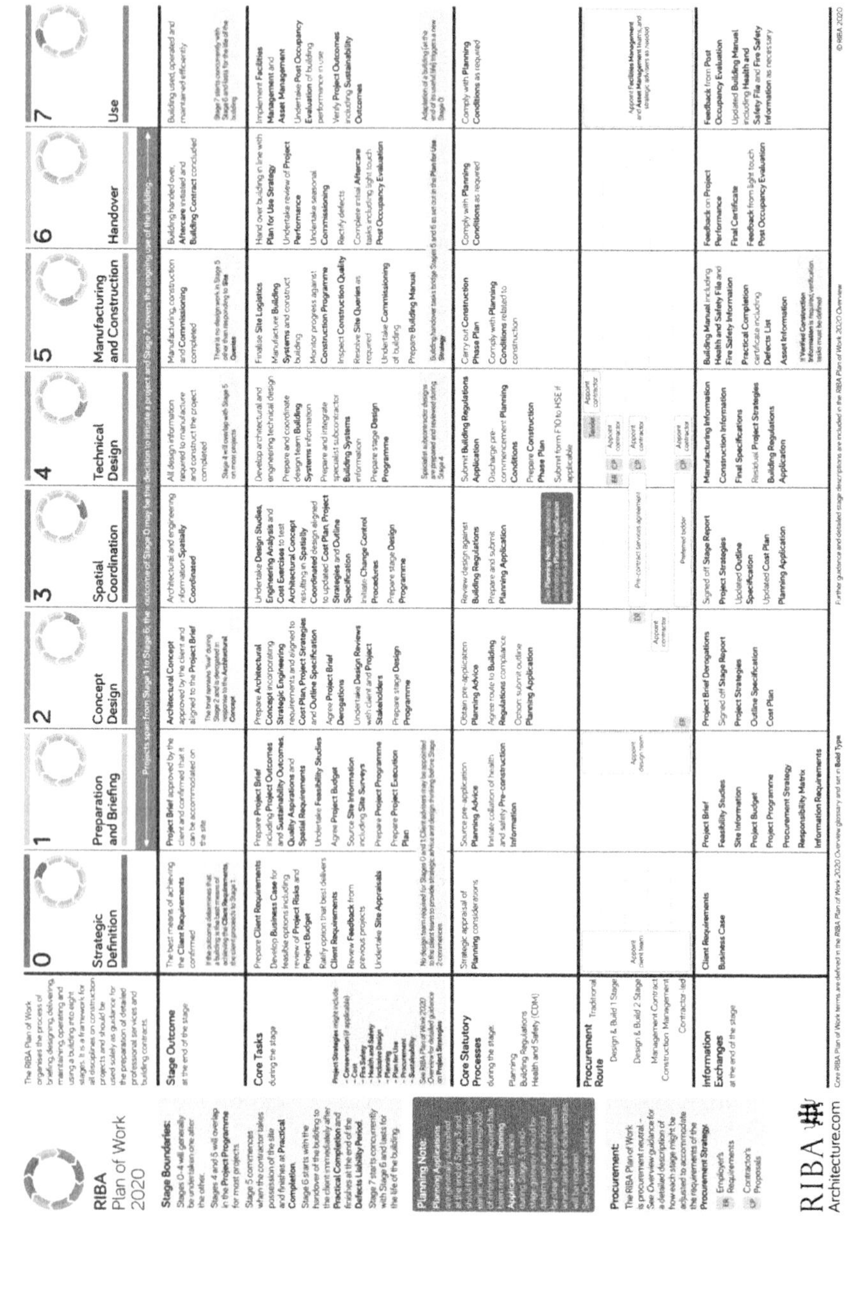

40 Dan Gerrella

Figure 3.1 RIBA Plan of Works 2020.

> **Box 3.1 The Architects Registration Board (ARB)**
>
> The Architects Act 1997 established the Architects Registration Board (ARB) as a statutory body. It is responsible for:
>
> - Prescribing the qualifications required to become an architect
> - Controlling the use of the title 'Architect'
> - Maintaining the UK Register of Architects
> - Regulating the profession through a Code of Conduct
> - Promoting standards of professional behaviour and competence
>
> The Code of Conduct establishes 12 standards for architects. Two specifically relate to communications:
>
> - Promote your services honestly and responsibly
> - Maintain the reputation of architects
>
> Organisations that do not follow these standards will be fined, suffer significant reputational damage and can be struck off the register.
>
> Source: http://www.arb.org.uk

Who are the audiences architects need to speak to?

Architects want to secure approval for their projects from peers and the community. Communications often need to target a broad audience: existing and potential clients, supply chain partners and relevant members of the public.

Messaging should focus on design quality, breadth of experience and community outcomes. These will be the most important influencing factors behind any appointment.

A positive reputation is important, as the chart demonstrates (Box 3.2). Half of all work is secured through direct appointments, so practices need to be recognised by name.

> **Box 3.2 Common ways of appointing an architect**
>
> Direct appointment – 50%
> Competitive fee bid or financial tender only – 21%
> Framework agreement with or without further competition for specific projects – 10%

> Invited competitive interview (no pre-qualification questionnaire PQQ) – 4%
> Expression of interest / PQQ only (no design work) – 3%
> Expression of interest / PQQ followed by competitive interview (no design work) – 3%
> Expression of interest / PQQ followed by design competition – 2%
> Invited design competition (no PQQ) – 1%
> Open design competition – 1%
> Other – 4%
> Source: RIBA Journal[3]

It is important that when an architect bids for work they have an established reputation. This will reassure the purchaser when they are making their appointment. This may influence everything from the ability to secure planning permission through to the final value and ability to sell the completed scheme.

Looking at the three types of practices outlined earlier, there are some broad differences in terms of how work is won:

- A **signature** architect is often approached because of who they are. Led by a well-known figurehead or team, the name is enough to convey the style of design and outcome that will be delivered.
- **Specialist** architects rely heavily on their track record to demonstrate how they are best placed to provide design solutions in their niche. They often work with others as part of a larger team.
- For **commercial** architects, there tends to be more competition. Their struggle is to avoid becoming commoditised as fees continue to be squeezed. Many tender processes give lowest price higher weighting.

Despite the differences between the three types of architect, they do target similar audiences. The differences are mainly driven by the approach the client wants to take and their budget. Hiring a 'name' may provide more value from a marketing and sales perspective, but it will require a larger investment in design fees. The key audiences to consider are shown in Table 3.1.

A strategic approach to PR when targeting these audiences helps to secure work and to gain support for schemes. Each audience segment should be researched, looking at aspects including their media preferences, issues that affect their ability to do business and their perceptions of the practice. This helps to understand how to tailor messages and content appropriately.

Beyond these specific areas, there are wider issues that matter to audiences. Broadly speaking, they cover societal issues, such as environmental, technological and economic challenges. Therefore, a broader general public may need to be

Communications for architecture 43

Table 3.1 Typical target audiences for architects' communications.

Local planning authorities	Local planning authorities are the planning department of the local borough, district or unitary council. All proposals for new development need to involve them.* They are responsible for developing the local plan which sets guidelines for development. Masterplanners can support this phase, developing the vision for the area. Architects may be involved in setting design and quality standards. LPAs also determine planning applications, looking at factors ranging from scheme quality to social impact. Local MPs, mayors leading combined authorities and councillors will be key stakeholders; their support for a scheme will carry weight with the planners and they can help to mobilise the community. (*note: for some areas, there may be others to work with instead, such as a non-metropolitan county council, mayoral development corporation or body such as a national park authority).
Developers	Developers are responsible for developing a building, either from scratch or in improving an existing asset. Developers benefit from support from architects in many areas – from feasibility through to delivery (see RIBA stages earlier) and as such are one of the main target audiences for architects.
Referrers	Other consultants such as quantity surveyors, engineers and specialist advisors are important as the developer client may have existing relationships with them. These audiences offer another potential route to winning a contract by instigating partnerships or recommending practices.
Banks / other funders	Potential funders of architectural projects will seek the lowest possible risk profile, so a practice with a good reputation may be preferred. This is because reputation may correspond to better delivery or a higher potential for sales.
Government	The UK government is responsible for regulating the industry and setting guidelines. It is a potential client for strategic work, as well as the construction and improvement of its own built assets. Policy decisions at this level will impact on architectural work – for example, the UK's commitment to be carbon net zero by 2050'.
Institutional investors	Institutional investors manage large portfolios of buildings and will often seek advice on improving asset value. This enhances the value of their portfolio, delivering better returns for their investors.
Built asset owners and managers	Assets regularly need improvement and maintenance. Architects support by advising on refurbishment, regeneration and adaptation.
Contractors	Contractors carry out construction works. There can be conflict between what the architect designs and how the contractor delivers this on site. It is important both work together effectively. A good working relationship reduces potential friction during construction.
One-off purchasers	This is a wide-ranging category and as such can be difficult to reach. Frequently this is achieved through the specific media aligned with the target audience, for example publications focused on luxury homes.

communicated with, depending on the issue and how architecture can have an influence.

> **Box 3.3 Ask the audience**
>
> When researching potential customers and their issues, it is important to speak to existing clients. Service reviews are a great way of gaining greater audience understanding and offer an opportunity to move the discussion away from projects. Areas to review include:
>
> - The client's business objectives and strategy
> - Future opportunities they will have for new work
> - Background on the practice (are they aware of any other relevant support that can be provided?)
> - Feedback on the existing/completed project, to identify areas for improvement
>
> The findings can be used to improve performance and within external communications as a way of demonstrating customer satisfaction.

Working with the media

Once the audiences are established, a plan on how to target them can be developed. One of the main ways is through the media. Common areas of focus within a media relations plan include:

- Increasing a practice's profile and brand – through corporate announcements and project news
- Attracting high calibre employees – through an employee-focused brand that describes corporate culture, progression opportunities and quality of projects and opportunities
- Establishing a voice in the industry relating to key trends – through thought leadership pieces, commentary and research

Architects are often concerned with the perceptions of their peers. Colleagues and competitors often review their work through design reviews and design panels. There are also specialist architecture critics working within the media. Practices therefore tend to focus efforts on promotion via the architectural trade press and specialist reporters. This does have value, as it reaches the audiences mentioned, and it can help attract new recruits and motivate existing staff.

However, a more successful approach in terms of winning work is to target sector-based media relevant to customers and prospects. For example, a practice that specialises in designing schools is advised to focus efforts on the education trade media. This increases the opportunity for new prospects to hear about the practice and acts as a good relationship enhancer with existing clients who consequently gain credibility in the eyes of their peers.

To help with planning, prioritise media into different categories such as core titles, specialist titles and 'other'. The latter still have some value but tend to be less relevant for key audiences.

What do journalists want?

Specific 'milestones' have a high likelihood of securing coverage:

- Project announcement, including design concept
- The scheme securing planning permission
- Any special features or innovation within the design, materials or process
- Project completion

These are broadly in order of priority. The architectural trade media, for example, will focus heavily on the announcement, analysing the design and conceptual approach to the building. When it comes to completion, there is likely to be less coverage, as the architectural story has already been told. That is, unless it is a high profile, prestigious building. If that is the case, investment in photography and an offer of a project tour improve the chances of coverage.

It should be noted that any element of controversy is likely to be picked up at some point during the project. For example, a project involving demolition or refurbishment of an old building often has some element of opposition. Some of this may be from local individuals and some may be from organisations, such as heritage groups. Either way, it will be enough to spark further media interest in the project. This could run and run, even beyond project completion, if the scheme still goes ahead, as legal challenges are sometimes raised.

The project team is often aware of these issues in advance, so communications teams need good relationships with them. This helps to manage the crisis better, as key people already know each other and can prepare statements as part of a crisis communications planning exercise.

The response should emphasise any mitigating factors influencing the decisions. Keeping with the demolition example, is it a commercial reason (the only way to get the scheme to work is to replace it with something else), health and safety related (the building is structurally unsafe or includes unsafe materials such as asbestos) or environmentally driven (replacing a poorly performing building with something new)?

Clients often want to see coverage relating to project milestones. Within the architectural media, these tend to be less desired. A practical completion story may get limited coverage, but in all likelihood will not be of interest. Milestone-based content is instead useful for owned social media channels and website. This is covered in the section 'The digital world'.

Structuring stories

While corporate news has its place (appointments, initiatives, financial results and the like), the majority of media relations work in architecture focuses on project news (in terms of regular news, see 'How to get the practice to stand out' for content marketing related to thought leadership).

Table 3.2 provides an example of a traditional structure for a project win press release.

This is by no means the only way of structuring a story but is included as a basic example. An architectural title would include the name of the architect much higher up in the release, if not in the introduction, and would focus on the design perspective and inspiration, the community benefits, the materials and the site context.

The determining factors will always be the target audience and where the piece is appearing.

Imagery

New schemes are usually presented with a computer generated image (CGI) of the final development. This is aspirational; buildings framed by sunny days and happy people walking around. Often the immediate context is removed or faded out, leaving the proposed building as the main point of focus.

CGIs vary in quality and style. Inevitably, there is a direct correlation between cost and quality. This can create issues. One of the purposes of a CGI is to build support for a scheme and promote it for sales purposes. There are many examples of cheaper CGIs being used because sufficient budget has not been allocated. It can create opposition, as people react to the image rather than the scheme description.

These initial phases show what people will expect to see when the project is finished. If there is a huge variance between the CGI and reality, expect negativity. This damages the practice's reputation because it creates a lack of trust in what is being delivered and leads people to believe they have been conned into approving something unrealistic.

Once this has happened, it is usually pointless to update the imagery to a higher standard. The damage has been done in terms of perception.

There are many reasons why the final product might not match up with the original imagery. Some value engineering may have occurred to ensure the project remained commercially viable, there may have been poor quality finishes when it was delivered by contractors or materials may have been changed due to availability issues. Whatever the reasons, it is important to manage expectations to avoid a backlash.

This also ties into the architectural codes of conduct discussed previously as well as those that exist within public relations and marketing. Communications should be transparent and honest. This is especially true when companies are seeking investment or advance purchases of their product (e.g. 'buying off plan' in the residential sector).

Table 3.2 Content for a generic project announcement (for construction trade media).

Section	Content ideas	Example
Introduction – paragraph 1	• What is the core outcome that will be delivered? (This should be a community benefit.) • What is the investment? • Where is it? • Innovative features?	Proposals have been shared for a £50m residential development that will create 200 new homes in Birmingham city centre.
Paragraph 2	• Who is delivering the scheme? • What else does it include? (Expand on details within the introduction)	Delivered by Midlands Developers, the scheme comprises a mix of one and two-bedroom apartments, complemented by ground floor retail and leisure.
Paragraph 3	• Further location detail and benefits	The site is located close to Moor Street station, offering good transport links within the city.
Paragraph 4	• First quote – person delivering the scheme	'We're hoping to attract young professionals who want to live within the city centre, with easy access to work and leisure…'
Paragraph 5	• Detail behind architectural approach	According to architects ANOther, the design of the scheme has been inspired by the area's industrial past.
Paragraph 6	• Second quote – architects, explaining design approach and why it is important	'The scheme is located on the site of a former metal works. We wanted to reflect that in the choice of our materials, from the brushed aluminium façade through to the choice of public artworks within the green spaces. 'It's an important scheme for the area, providing much needed…'
Paragraph 7	• Any relevant further context – other developments in the area, similar projects from the architect and developer to demonstrate track record	It is the latest project secured by ANOther architects in the area, following the recent appointment to masterplan the redevelopment of an old retail park in Selly Oak.
Paragraph 8	• Final details for the release – next steps, e.g. when will it start on site/go in for planning/etc.	Subject to planning consent, the scheme will start on site late 2020, with completion expected in summer 2022.

Beyond CGIs

Different media have different preferences, but anything that shows the design process and the proposed scheme will be useful; if the target publication is not interested, it will still be useful for the practice's content channels (see 'Telling the story' for more on this).

Other examples of imagery include:

- Artists' impressions, such as sketches, watercolours or detailed drawings
- Photography of models, ranging in detail from abstract massing models (showing the relative size of the building) to models that include fine detailing
- Precedents and examples taken from other projects to show what has inspired the new scheme and how it might look when completed
- Elevations, outlining the measurements and scale of the development
- Site plans and technical drawings

When the project is completed, final photography can be commissioned. Like the types of practice identified in the introduction, there are various approaches to photography, with some practices using 'named' photographers that they fly all over the world to take photos of their schemes. A healthy photography budget is a must.

The architectural media tend to favour shots of the building, both during the day and at night, with clear lines and detailing. Often, these photos will not include people – the building is presented as a piece of art, standing on its own terms. However, 'in use' shots should also be taken, as they communicate scale, activity and a sense of place.

How to get the practice to stand out

Like many consultancy based services, architecture has barriers to entry in the form of accreditations and qualifications (as explained in the Introduction). In many ways, this creates a level playing field when it comes to architects selling services. Most offer largely the same skills, to the same standards, for roughly the same price. How then can a practice be distinguished?

Messaging should avoid focusing only on what the practice does (products/services) and the business structure that allows it, e.g. 'We have 10 studios, with over 200 staff, offering architecture services'. Clients want to know that these foundations are in place, but they are not the determining factor when it comes to making a buying decision.

In professional services, people are the most important aspect. Clients want to know about the individuals they will be working with day to day. This is especially true in architecture and urbanism, as contracts can last anything from two to twenty years. It is important that a positive working relationship can be built.

The other strand to focus on in messaging is the 'why' behind what the practice does. What is the ultimate purpose and desired outcome of the practice's work? How does the practice's culture drive that purpose? In architecture, this is often tied to the concept of placemaking: delivering places that reflect their local context and the needs of the community. This links in with the importance of delivering social value (see section on social value below).

Becoming a trusted advisor

With the above in mind, it is important to position key individuals within a practice as experts in their core sectors, service lines or niche areas of knowledge. This can be achieved partly through positioning in the media via comment opportunities. These can be based on trends in the sector or in direct response to something topical within the news agenda.

Detailed case studies, research and stakeholder engagement all offer routes to developing this insight. This helps position a practice effectively and increases the likelihood of winning work.

This ability to demonstrate knowledge is one of the things that will determine whether a potential client buys from a practice – along with track record, market knowledge and ability to provide relevant services in the right places.

In addition, clients want to work with an advisor they can trust, with the right skills and knowledge to support their business aims. Practices should promote the personalities of staff as much as their design style and creative approach.

A further way of enhancing relationships with clients is to involve them in communications. Help them to tell their stories about the places the practice has designed for them. It is more valuable to hear their perspectives, as it provides an endorsement for the project and a more accurate reflection on the design outcomes and value.

Telling the story

One of the advantages of working in PR in this sector is that you are there right at the beginning of a project. The architect is the one taking the brief from the client, researching the local context and developing an idea from an abstract sketch through to a detailed design. This provides an opportunity to tell a story.

Good contextual design considers an area's history and culture, among other priorities. Such references offer both a chance to look to the past and also to consider an optimistic future. How will the building or masterplan benefit the community? What is it about the design that will make this place stand out and attract people? Are there any wider social issues that will be solved or improved as a result of this development?

Architects can take some of this process for granted, seeing it as 'the way things are done'. Communications professionals must draw these themes out, working to discover the passion and nuances that went into the design, as well as the desired outcomes that the project will deliver.

For a more rounded piece, involve others within the project. If it has secured planning consent, try talking to a local councillors or a nearby community group or business that supports the proposals. If it is at the concept stage, ask the developer or funder why the scheme has potential, what their aim is and why the practice was selected.

Drawing from a range of people offers more context for the project beyond the architecture. It also starts to build relationships, which may pay-off in terms of any follow-up PR activity later. This includes obtaining testimonials or creating case studies later down the line.

The digital world

Broadly speaking, the architectural sector is quite traditional in its approach to business and marketing. Nowhere is this more apparent than in the industry's struggle to adapt to new technology.

In 2018, RIBA teamed up with Microsoft to develop a report on digital transformation.[5] It outlined benefits from technology including the ability to collaborate, innovate and improve productivity. This mainly focused on Building Information Modelling (BIM) and the Internet of Things (IoT).

Nine years after the introduction of BIM in the UK as part of the Government's Construction Strategy and targets for public-sector projects,[6] research by NBS shows that, despite the government mandate and extensive positive publicity about the benefits, the latest adoption rate is just 69%.[7]

This reluctance to innovate seems to extend to communications tactics. For example, the varied usage of social media between different practices shows that there are still significant improvements that can be made. In 2020, I conducted a review of the social media channels used by the AJ100's top 20 practices which revealed the following proportion of practices with an account:

- Twitter: 85%
- Instagram: 55%
- YouTube: 25%

Analysis from Hootsuite found that 75% of B2B businesses use YouTube, so architecture is well behind the average.[8] Further research has shown that 68% of people would prefer to learn about a new product or service by watching a short video, compared with 15% for written content.[9] Video clearly has potential for the sector.

The review of the top 20 practices also showed that there is a reliance on corporate social media channels rather than personal channels. For example, only five of the principals or managing directors use Twitter, with three having fewer than 100 followers.

Anecdotally, during my work with architectural practices, social media was often seen as irrelevant by the senior team. That is, except for Instagram. As mentioned earlier, architects are highly creative, visual people. The channel is perfect for what interests them; they can find arty shots of buildings, close-ups of detailing and imagery offering intriguing colour palettes or materials. To a lesser extent, Pinterest works in much the same way.

Social media also helps architects to build their own profile and personality online. This is useful for clients, offering an extra touch point and helping to develop relationships. As mentioned above ('Becoming a trusted advisor'), these personal interactions are important influences within the purchasing process and project delivery phase. The lack of practice leaders on social media could therefore be considered a risk.

Another way to improve social media engagement within a practice is to demystify it. Where possible, show the results. Social media can be great for lead generation, and through in-platform analytics, link tracking and website tracking, it is possible to follow a user's referral path through to a predefined action or 'event' (such as downloading a report or completing an enquiry form). The capacity of social media to achieve this should be communicated widely throughout the practice.

Other tips for social media adoption include:

- Encourage staff to use channels that already interest them.
- Establish the basics with some introductory training for key channels.
- Clarify any rules in a social media policy – people can be discouraged from using social media because they are worried about what they can and cannot share.
- Ask staff to engage with corporate content (it offers them content to reuse easily).
- Do not be too prescriptive about what staff can do (beyond the policy).
- Share great examples from within the practice to inspire people.

Emphasise that content can demonstrate the design process. In architecture, there is the tendency to want to wait for the perfect shot of the building at completion. Social media should be more 'in the moment'. Most of it is disposable content. For example, a tweet will gain most of its engagement within the first 24 minutes.[10] It is designed to capture interest at a point in time. Further content can come later.

This is also true of video. Mix on-site footage with the more corporate, professional approach. To capture a sense of place, people and activity-based footage is needed. Mobile phone video quality is high enough that live streamed content can look great. For a small investment in a microphone and tripod, a phone can be used to film in a more traditional style. Try to encourage architects to do this. They regularly visit site and will be able to provide footage that explores what the building looks like. This lower budget approach provides a more realistic portrayal.

Another useful tactic is e-newsletters, as they offer a way of communicating directly with clients. The architecture industry is one of the best performing in terms of open rates (23%) and click-through rates (2.7%).[11] By providing audiences with the right content at the right time, this can be highly successful.

The importance of track record

An established practice should have a track record of successfully delivered projects. Case studies, testimonials, project information sheets and high-quality completion photography are all useful tools as part of the selling process. They can be used to support bids, award entries and speaker slot pitches.

Content should explore the strategic decisions that were taken, the approach to design, the inspiration for the creative process, and how the project was delivered. This should focus on client service, design quality and final outcomes. Aside from the marketing benefits of doing this, the information can be fed back into the business, reinforcing positive behaviours and driving continual improvement by reflecting on completed work.

Awards

Architectural awards are often voted for by peers or the general public and are a good indication of affection for the final product. The most important are those that will enhance a practice's reputation in the minds of their key audiences.

Most criteria require basic facts and figures on the project. This will usually include key dates (planning permission awarded, start on-site, completion), project value, services provided, size and location.

Questions are often asked about accreditations and any performance measures, especially if sustainability focused (see Chapter 11).

Beyond this, judges will be looking for the design narrative from the concept onwards. This links back to the earlier points on what the media is looking for.

Awards programmes are often supported by at least one media partner, leading to promotional opportunities if practices or projects are shortlisted, and even more so if they win.

However, in the case of most media-sponsored awards, the resulting media coverage can be limited to the specific publication running the awards. In such cases, social media provides an opportunity for the communications professional to extend the promotion of the win.

Social value

Architects play a key role in society, designing the form and function of the built environment. But they require permission to do it, provided by communities, local authorities and via industry regulation.

An important aspect of achieving this is through public engagement. This process involves raising awareness of projects and communicating the benefits.

One of the outcomes of good development is creating social value. There are many definitions of social value, but Bristol City Council's is one of the most succinct: 'Social value is about maximising the impact of public expenditure to get the best possible outcomes and recognising that local people are central to determining how these can be achieved'.[12]

It is a legal requirement under the Social Value Act 2012 that public sector organisations consider social value as part of their procurement process.

Social value is also important from an ethical perspective, for all projects. The construction process often causes much disruption to communities.

There is an increasing expectation for companies to 'do good'. Consequently, both reputation and ability to do business depends on it. The UK Green Building Council has stated, 'We are experiencing a momentous shift in our expectations of business, with organisations under increasing pressure to demonstrate their contribution to society'.[13]

While social value is often thought of in terms of a developer's responsibility, architects have a role too. They tend to be involved throughout the process and are key influencers when it comes to determining the final outcomes of a project.

Communicators can play a strategic role in this process by using their core skills to:

- Develop an understanding of the local context, such as key economic and demographic factors, through horizon scanning exercises.
- Identify the positive outcomes that are most desired by the community though stakeholder engagement via detailed two-way dialogue.
- Engage communities and develop a sense of ownership of both the proposals and end product through sustained communications.

The more extensive the engagement, the more chance there is of delivering better community outcomes. A shared vision between the professional team and the local community setting the goals and desired legacy for the project can be created and agreed. This can include quality factors, practical uses and a top level approach to the problems that the project is aiming to solve.

These desired outcomes can be arranged under three broad themes: jobs and economic growth, health, wellbeing and the environment and strength of community. These three themes should be considered both as part of the delivery phase (creating jobs in the construction of the project) and following completion (spaces created for new jobs, such as office or retail positions).

In turn, this will lead to a positive reputation for both the development and the wider construction industry.

That improved reputation is something that the whole industry can benefit from, with skills shortages, lack of diversity and the challenges of securing planning being just three of the issues that can benefit.

Selling the vision

Stakeholder engagement is an important part of an architect's role, both in terms of developing concepts and securing planning.

Looking at the former first, it involves communicating with the end users of the project being delivered. This is where a PR professional should use their research skills to identify the right people to speak to so that they can be involved in the process.

Ask stakeholders what they want and need from the project. Their expertise and knowledge will vary, so do not focus on the technical side. Explain things so that they can understand the concepts involved. A good way of doing this is through interactive workshops that explore key ideas before asking them for their views (see Box 3.4).

Box 3.4 A case study in stakeholder engagement from Broadway Malyan

Density in urban design has negative connotations. Many think of high density as tall skyscrapers packed closely together. Concerns are often raised about visual impact and the effect on local services and infrastructure.

When talking to communities then, it is important to develop their understanding of what density means in practice.

Broadway Malyan had this challenge in Dar es Salaam, Tanzania when working on a strategy for the area around a new Bus Rapid Transit (BRT) line.

A creative solution was needed to overcome the stereotypical views about density. Knowledge gaps had to be filled, and cultural and language barriers had to be considered.

The answer was to use LEGO to represent the volume of planned development. Different coloured bricks symbolised different building uses, and an initial skyline was proposed. Participants were then encouraged to take pieces apart and reconfigure them to see how they could balance the different uses.

This tactile approach made it easier to discuss the concepts of a liveable city. The familiarity of the bricks acted as an ice-breaker and helped build a two-way dialogue between the planners and the community. By the end, they understood the challenges of making everything fit in the land available. They also saw how high density could be positive, as long as it was appropriate for the local context.

The World Bank was one of the funders behind the project. The workshop was so successful that they invited Broadway Malyan to run it again with their team in Washington.

Chyi-Yun Huang is a senior urban development specialist at the World Bank. Commenting on the approach, she said:

> Interactive, hands-on activities are much more effective in terms of engagement and messaging. These help bridge the gap between the speakers and the audience and most importantly bring down some false common beliefs like those identifying density with a forest of tall skyscrapers. It was critical to clarify this point in Dar es Salaam.

Public consultations

A statutory part of the planning process is demonstrating that the proposals have been effectively communicated to the public. To help communicators plan and deliver a successful public consultation, I have developed the 'Four As Model'. This model describes the three core elements that should underpin all messaging during the process, with the end goal of achieving advocacy from the community (see Table 3.3).

The issue of accessibility is especially important. Consultations should target as many people within the community as possible to accurately reflect local needs. This will include hosting an exhibition so that local people can come and view the plans and comment. People will want to know 'what's in it for them' and how their views will be considered and fed into any final design.

When explaining the scheme, transparency is important. Transparency features in the code of conduct that architects work to (see Introduction) and as part of the rules within communications organisations such as the Chartered Institute of Public Relations (CIPR)[14] and the Chartered Institute of Marketing (CIM).[15]

Tactics for consultation are detailed within Chapter 2 (Communications for planning). The Local Government Association website also has more information.[16]

Outside of consultation events, regular communication is a must even if there is not much to say. This will keep the community engaged with the process.

Table 3.3 The 'Four As Model'

Accessible	+ Aspirational	+ Attainable	= Advocacy
Consultation content is easy to understand (no jargon). A variety of channels are used to invite feedback. Different communications tactics are used to reflect the ways people absorb information. Diverse voices of the community are welcomed and heard.	The community's needs are identified and agreed through structured dialogue. A vision is developed in partnership with the community focused on their desired outcomes.	The vision is achievable and takes into account any site, political or commercial constraints. The final product will be a reflection of the concepts and proposals shared.	Community will support the scheme through planning. Supporters will positively represent the project to media. Community will defend the project's interests.

Working internationally

For international practices, there are two models: 'fly in, fly out' or open a local studio. For the former, this may include setting up short-term partnerships, working in client or temporary offices or travelling to the location as required. The advantage is that it is lower risk than setting up a new business in country. However, from a brand perspective it is often viewed negatively in the local market.

One of the tenets of architecture is understanding the local context of place. Culture, history, demographics, climate, environment and social needs are all factors that should be considered as part of the design process. The best way to understand this is to embed people in the location and work with local staff. Not to do so, in Asia for example, runs the risk of being viewed as imperialistic: there just to exploit the opportunities.

Messaging has to consider this carefully. How will the practice commit to the local environment that it is working in? Is there a legacy that will be delivered beyond the project, for example upskilling the local workforce?

Messaging can also describe existing best practice and how this will benefit the new country being entered. UK credentials and qualifications carry weight internationally, so emphasising these standards and accreditations can be of benefit.

Working with media varies from country to country. Take the time to investigate and learn about timings, procedures and desired content. Speak to clients and staff about the media that they value to inform the development of targets and contact lists. Also, invest time in getting copy translated into the local language. Do note that translators are not perfect. It is much better to find a freelancer in the country who can translate and localise content. This will ensure that the language used is appropriate and reflective of local dialects and phrasing. They will also have their own relationships with media that will be of benefit.

Issues facing the sector

Architecture is one of the sectors most influenced by wider societal change. After all, it involves the design of the places where we all live, work and relax.

Areas of note currently include:

- **Urbanisation:** by 2050, 68% of the world's population will be in living in cities.[17] Cities will replace countries as the main drivers of economic growth. Therefore, cities must attract highly skilled, talented people. This depends on having the best infrastructure, opportunities and quality of life. The scale of development required in some countries will lead to resource challenges globally in terms of skills and materials.
- **Demographic changes:** people are living longer, meaning that buildings and cities need to cater more for health and wellbeing needs.[18] Workplaces can be home to as many as five generations, each with its own style of working.

Residential communities need to focus on accessibility and facilities. Parts of the world, such as India and China, are seeing a growing middle-class, changing economic models.
- **Climate change:** scientists are predicting the 'point of no return' in the next ten to fifteen years.[19] The built environment is responsible for approximately 36% of the world's carbon emissions, including construction and in use.[20] Resilience is also an issue, with extremes in weather increasing. In June 2019, RIBA declared a climate emergency, saying it was the biggest issue affecting the profession.[21]
- **Quality and safety of the built environment:** in the UK, perceptions of the final construction product are low, with poor quality workmanship in new-build developments, fire safety issues and project delays and overspends just three of the issues regularly featured in the media.
- **Modernisation of construction:** new materials and methods of construction exist and are being used at varying levels across the industry. New technologies, including robotics and artificial intelligence (AI), will challenge the status quo. Many reports have argued that the industry needs to update the way things are done before it is too late. 'Modernise or Die: The Farmer Review of the UK Construction Labour Model' addresses these topics.[22]

The architecture industry is also facing its own specific challenges. Fees are being squeezed and new challengers are entering the market, as technology companies investigate opportunities. There is an issue around productivity, with gaps between client demands and what is delivered, and the political and planning environment in the UK is regularly criticised.

The process of gathering client feedback and 'in-use' analytics to check building performance against concepts is sporadic, and there is limited consensus on what 'good' looks like from a design perspective.

The wider architecture sector seems to be in denial about some of these problems, and there are underlying issues such as a lack of diversity and gender balance.

Tackling these problems offers a chance for practices to stand out from competitors and enhance their reputations. This will help attract and retain staff and customers, with people increasingly driven by ethical considerations.

Box 3.5 An industry that struggles to retain women

Since April 2017, companies with more than 250 staff are legally obliged to publish information regarding their gender pay gap.[23] However, a huge number of practices (including large international practices, which are not required to include non-UK staff within their totals) fall below this threshold.

> Architects Journal (AJ) is one of the UK's leading trade publications in the sector. In 2012, it started a campaign which included an industry survey, events, awards and a partner programme focused on women in architecture.[24]
>
> The 2019 survey revealed a pay gap at every level.[25] The worst level was associate director with a gap of 12.5%.[26] Motherhood is often cited as a reason for the pay gap, with the short career break due to maternity leave used to justify the divergence. However, the same survey revealed a gap at entry level (2.8%), somewhat disproving this.
>
> With a huge impact on reputation, staff retention and recruitment, an architectural practice that focuses on solving this issue could make significant reputational gains, assuming that any messaging in this area is backed up with direct action and genuine business improvements.

Conclusion

This chapter outlines some of the key challenges and successful communications tactics within the architecture sector. These are by no means exhaustive but should provide a good foundation.

The best tip is for communicators to immerse themselves in the sector. Read the trade media. Speak to architects. Attend conferences and events. Learn about the language and processes architects use, and the issues they face. Do this, and those working in the sector will be able to advise architects strategically and successfully on what they need to do to ensure that their projects and practices prosper.

Notes

1. RIBA Plan of Works [RIBA] https://www.architecture.com/knowledge-and-resources/resources-landing-page/riba-plan-of-work [Accessed 29 June 2019].
2. Pathways to qualify as an architect [RIBA] https://www.architecture.com/education-cpd-and-careers/how-to-become-an-architect [Accessed 29 June 2019].
3. What's the best way to win work, and how do you avoid giving too much for free? [RIBA Journal] February 2014: https://www.ribaj.com/intelligence/investment-strategy [Accessed 29 June 2019].
4. Climate change: UK government to commit to 2050 target [BBC News] 12 June 2019: https://www.bbc.co.uk/news/science-environment-48596775 [Accessed 29 June 2019].
5. Digital transformation in architecture [RIBA, Microsoft] 5 July 2018: https://www.architecture.com/-/media/gathercontent/digital-transformation-in-architecture/additional-documents/microsoftribadigitaltransformationreportfinal180629pdf.pdf [Accessed 29 June 2019].

6 Government construction strategy [Cabinet Office] May 2011: https://assets.publishing.service.gov.uk/government/uploads/system/uploads/attachment_data/file/61152/Government-Construction-Strategy_0.pdf [Accessed 29 June 2019].
7 National BIM report [NBS] 17 May 2019: https://www.thenbs.com/knowledge/national-bim-report-2019 [Accessed 29 June 2019].
8 28 Twitter statistics all marketers need to know in 2019 [Hootsuite]: https://blog.hootsuite.com/twitter-statistics/ [Accessed 7 August 2019].
9 Video marketing statistics 2019: the state of video marketing 2019 [Wyzowl]: https://www.wyzowl.com/video-marketing-statistics-2019/ [Accessed 7 August 2019].
10 28 Twitter statistics all marketers need to know in 2019 [Hootsuite]: https://blog.hootsuite.com/twitter-statistics/ [Accessed 7 August 2019].
11 Email marketing benchmarks [Mailchimp]: https://mailchimp.com/resources/email-marketing-benchmarks/ [Accessed 29 June 2019].
12 Social value in new development: an introductory guide for local authorities and development teams [UKGBC] March 2018: https://www.ukgbc.org/wp-content/uploads/2018/03/Social-Value.pdf [Accessed 29 June 2019].
13 Social value in new development: an introductory guide for local authorities and development teams [UK GBC] March 2018: https://www.ukgbc.org/wp-content/uploads/2018/03/Social-Value.pdf [Accessed 29 June 2019].
14 CIPR Code of Conduct [CIPR] https://www.cipr.co.uk/content/members/public-relations-register-overview/cipr-code-conduct [Accessed 29 June 2019].
15 CIM Code of Professional Conduct [CIM] https://www.cim.co.uk/media/1542/code-of-professional-conduct.pdf [Accessed 29 June 2019].
16 Consulting residents [Local Government Association] https://www.local.gov.uk/our-support/guidance-and-resources/comms-hub-communications-support/resident-communications-4 [Accessed 29 June 2019].
17 2018 Revision of World Urbanization Prospects [United Nations] 16 May 2018: https://www.un.org/development/desa/publications/2018-revision-of-world-urbanization-prospects.html [Accessed 29 June 2019].
18 Living longer: caring in later working life [ONS] March 2019: https://www.ons.gov.uk/peoplepopulationandcommunity/birthsdeathsandmarriages/ageing/articles/livinglongerhowourpopulationischangingandwhyitmatters/2019-03-15 [Accessed 29 June 2019].
19 Earth quickly heading for 'point of no return' unless we take immediate action, climate scientists warn [Independent] 30 August 2018: https://www.independent.co.uk/news/science/climate-change-global-warming-point-no-return-floods-a8515431.html [Accessed 29 June 2019].
20 2018 Global Status Report: Towards a zero-emission, efficient and resilient buildings and construction sector [Global Alliance for Buildings and Construction] June 2018: https://www.globalabc.org/uploads/media/default/0001/01/0bf694744862cf96252d4a402e1255fb6b79225e.pdf [Accessed 29 June 2019].
21 RIBA declares environment and climate emergency and commits to action plan [RIBA] 27 June 2019: https://www.architecture.com/knowledge-and-resources/knowledge-landing-page/riba-declares-environment-and-climate-emergency-and-commits-to-action-plan [Accessed 29 June 2019].
22 Construction labour market in the UK: Farmer review [Gov.UK] 17 October 2016: https://www.gov.uk/government/publications/construction-labour-market-in-the-uk-farmer-review [Accessed 29 June 2019].
23 Gender pay gap reporting: guidance for employers published [Gov.UK] 28 January 2017: https://www.gov.uk/government/news/gender-pay-gap-reporting [Accessed 29 June 2019].

24 Shock survey results as the AJ launches campaign to raise women architects' status [Architects Journal] 16 January 2012: https://www.architectsjournal.co.uk/news/daily-news/shock-survey-results-as-the-aj-launches-campaign-to-raise-women-architects-status/8624748.article [Accessed 26 June 2020].
25 Welcome to Women in Architecture [Architects Journal]: https://www.architectsjournal.co.uk/women-in-architecture/about [Accessed 29 June 2019].
26 Women in architecture: mind the pay gap [Architects Journal] 14 February 2019: https://www.architectsjournal.co.uk/news/women-in-architecture-mind-the-pay-gap/10039915.article?blocktitle=survey&contentID=22614 [Accessed 29 June 2019].

4 Communications for major contractors

Embracing twenty-first-century communications challenges

Andrew Geldard

Introduction

UK construction is at a pivotal stage. The industry has been seriously blighted over several years by economic uncertainty caused by a protracted and at times acrimonious Brexit from the EU and then by the devastating impact of the COVID-19 global pandemic.

The current economic outlook is almost impossible to predict. After a record collapse in construction activity in April 2020,[1] followed by a sharp increase in July 2020 which staged its fastest jump since October 2015,[2] the industry is holding its breath, waiting to discover how severe the long-term ramifications of the pandemic will be.

All eyes are on the Construction Leadership Council's (CLC) 'Roadmap to Recovery'[3] – a strategy to drive recovery in the construction and built environment sectors, and through them, the wider UK economy, following the pandemic and economic downturn.

As the strategy states:

> The strategy aims to increase the level of activity across the construction ecosystem, accelerate the process of industry adjustment to the new normal, and build capacity in the industry to deliver strategic priorities, including: increasing prosperity across the UK; decarbonisation; modernisation through digital and manufacturing technologies; and delivering better, safer buildings... The outcomes will be a more capable, professional, productive and profitable sector, which delivers better value to clients, better performing infrastructure and buildings, and competes successfully in global markets. Failure to act will miss an opportunity to deliver this, and risks the industry lapsing into a longer term recession, which erodes capability and skills, and leaves a smaller, weaker sector as a legacy.

There are three phases to the plan, to be delivered over two years (2020–2022), although in reality they are not sequential and work is progressing across all three at once:

- Restart: increase output, maximise employment and minimise disruption (0–3 months).
- Reset: drive demand, increase productivity, strengthen capability in the supply chain (3–12 months).
- Reinvent: transform the industry, deliver better value, collaboration and partnership (12–24 months).

Encouragingly, in line with this strategy, much of the UK's political response to the disruption caused by coronavirus has been to 'build, build, build'.[4] Likewise, construction quickly adapted to new safety requirements created by the pandemic, and with the Government putting infrastructure at the heart of the UK's recovery, the industry is ready and eager to seize that opportunity.

Meantime, 'old problems' that have beset the industry have not gone away and will still be there for us to address over the coming years. In particular, the industry has a huge demographic challenge to contend with: fixing the outdated image that construction is all about 'men in hard hats digging holes', to give it the best chance of attracting a new generation of young people to meet the skills gap.

This is an urgent priority. According to the Construction Skills Network (CSN) report 2019–2023, published in February 2019 by the Construction Industry Training Board (CITB),[5] more than 168,000 new recruits will be needed to meet demand and replace those who are leaving construction (even before we consider more recent job losses). Reaching out to a new generation has never been more important, and communicators are at the forefront of moulding an image and reputation that will help do that.

Everyone involved in construction has a role to play in showcasing the dynamic, fast evolving sector that offers an exciting, varied career path with good financial rewards for those that thrive in it. Later in this chapter, we will look at how PR professionals in construction can harness the potential of people as publishers of content and ultimately become effective brand ambassadors.

Becoming an inclusive industry

According to the CITB, only 14.5% of construction workers are female, and the proportion of women working in skilled manual trades is a disconcertingly low 2%.[6] This gender imbalance is bad for business as well as a wasted opportunity; our industry is cut off from recruiting nearly 50% of the talent pool.

In 2015, a report by McKinsey, 'Delivering through Diversity',[7] said that companies in the top quartile for gender diversity on executive teams were 21% more likely to outperform on profitability and are 27% more likely to have superior value creation. Diverse teams perform better and make companies more profitable. Also, the relatively static gender balance data that never goes much beyond 12–14% of construction workers being women suggests an inability to retain and develop skilled women.

So what's holding construction back from cracking the gender equation?

It remains a hard task to convince people that a career in construction can deliver a rewarding future, especially women and those from minority backgrounds, who for years have not been targeted for careers in the industry and who now are left with little understanding of its opportunities. When parents want their children to go into accountancy, law or advertising, and careers advisors are not able to convince them otherwise, the onus is on the industry to be bold and imaginative in tackling this.

Rebalancing a male dominated industry to become more gender neutral is a long process but necessary to ensure sustainability of new talent. Willmott Dixon became the first contractor to deal with it head-on – in 2018, group chief executive Rick Willmott made the pledge in a YouTube film[8] that the company is targeting gender parity by 2030, citing that the very future of the industry can't be sustained if it's only 'fishing in 50% of the talent pool' and that the business needed to address this as a key priority.

> **Box 4.1 Case study: Willmott Dixon's drive for gender parity by 2030**
>
> Willmott Dixon's group chief executive Rick Willmott made a personal pledge to tackle the issue that only one in five of the company's workforce was female by committing to gender parity across all levels of management by 2030. The business case was made clear to staff: successful companies are proven to be those that provide a diverse and inclusive environment, where people feel challenged, contented and included within complementary teams.
>
> Achieving the gender parity aim would be done through culture change and by adopting new and innovative methods such as agile working to create a work environment that is inclusive for all parts of society.
>
> This was launched to employees via a film in which Rick talked about the business imperative and everyone's part in helping achieve the target. Also, a series of blogs by members of the senior management team on their own role in this process helped to underline the message, coupled with case studies of projects that were successfully adopting agile working.
>
> It was also launched to the wider industry through Willmott Dixon being the lead partner of *Construction News*' Inspire Me campaign, which sought to promote and encourage the conditions that would see more women take on more senior leadership roles throughout the construction industry. By partnering with the Inspire Me campaign, Rick Willmott's message was heard by thousands of people across the UK, both in the magazine and via a series of live event roadshows across the UK.
>
> By making it a very personal commitment, Rick Willmott became the figurehead and also proved to employees that he was serious about the issue,

64 *Andrew Geldard*

> that it would not be a short term 'management fad' but long-term and fundamental to business growth.
>
> The benchmark of how effective this has been came in late 2019 when the FT published its first ever list of the best 700 companies in Europe for diversity.[9] Companies appearing in the FT's Diversity Leaders 2020 list were identified in an independent survey of more than 80,000 employees across the countries and sectors covered, which included surveying the performance of 10,000 privately held and publicly listed companies employing at least 250 people.
>
> Willmott Dixon was the UK's highest placed company to appear in the list, coming third out of the 700 organisations ranked for their approach to inclusivity, ranked above household names including Ikea, Google, Tesla and Rolls-Royce.

Other construction companies have followed suit. In July 2019, Wates announced its intention to improve the gender balance of the workforce at all levels, with women accounting for 40% by 2025.[10]

The business priority now for PR professionals and communicators in construction is how they help their companies become more diverse and inclusive, where individuals feel valued and able to contribute. This means tackling stereotypes such as men wearing hard hats or carrying bricks, which reinforce the wrong perception that construction is only ever going to be male dominated and about working on sites.

This is about showing a sophisticated and innovative industry, with many different roles and disciplines that are not just site based, where people of all genders and backgrounds can achieve their career aspirations.

Doing this means challenging marketing that often resorts to type with rows of people in high-viz vests and hard hats appearing on pop-up stands at careers conventions. Communicators are in a position of influence to change this, and if that means picking it up with the local site or office team, then go for it!

Realising the potential of a dynamic sector

Yet the challenges don't end there for communicators.

A series of corporate failings has underlined how business models and strategies have not been robust enough to thrive in what is a competitive marketplace for small and medium-sized enterprises (SMEs) and major contractors. A combination of wafer-thin margins and bad business planning means it only takes a few miscalculations to cause severe upheaval that can undo years of careful planning and loyalty from a workforce.

A classic example of this was the demise of major contractor Carillion in January 2018. Wikipedia tells the story: Carillion was created in July 1999, following a demerger from Tarmac. It grew through a series of acquisitions to become the second largest construction company in the UK, was listed on the London Stock Exchange and in 2016 had some 43,000 employees (18,257 of them in the UK). Concerns about Carillion's debt situation were raised in 2015, and after the company experienced financial difficulties in 2017, it went into compulsory liquidation on 15 January 2018, the most drastic procedure in UK insolvency law, with liabilities of almost £7 billion.[11]

But in contrast with such high-profile failures, there are also outstanding successes and turnaround stories, such as the success of Balfour Beatty's 'Built to Last' initiative set by its chief executive in 2015 which saw the country's biggest contractor increase profit margins and improve productivity. The firm has now set up a 'New Normal Taskforce' which is generating hundreds of ideas from employees covering areas such as new markets and returning to work post-COVID.[12]

That's why working as a communicator in the built environment has never been more exciting. There are literally a multitude of ways to influence and shape the future, and it's somewhere where innovation gets you noticed.

In the twenty-first century – which is typified by high-tech devices and smartphones – the industry is going through profound change in order to compete. Much is already in progress, from off-site construction techniques and prefabrication of materials to the use of digital technologies like BIM, 3D printing and augmented reality (AR). Also, the application of drones is aiding productivity, while technologies such as robotics and AI are waiting for their moment. This also links back to attracting a new generation of thinkers who see the world through a different lens and have a unique opportunity to help bring disruption and innovation to the future of construction.

Purpose-driven values

A new post-Grenfell, post-Carillion era is also seeing the rise of a value-based culture where it's not just enough to win work and to build. Companies don't win respect by boasting about turnover targets and being 'one of the biggest' by the size of their order books.

Now an organisation needs to articulate how it is improving society's general wellbeing. This is not a box-ticking exercise; it's about how companies and their people leave a legacy of improving life chances and promote social mobility.

Research in 2019 by Willmott Dixon (carried out by its annual Your Say online survey among its own people)[13] showed that four out of five people want to work for a values-driven company where they can play a part in improving other people's lives. It is also important for the so-called Millennials that they work for a company that is aware of its commitment to society, which links to the earlier mention of how the industry attracts the next generation.

2019 was the year when big business recognised the importance of value-driven companies. In August 2019, a statement from The Business Roundtable,[14] a group representing chief executive officers of major US corporations, set out a new definition for the 'purpose of a corporation'. CEOs at companies including Apple, J.P. Morgan, Boeing and Amazon now believe that investing in their people, delivering value to customers, dealing ethically with suppliers and supporting outside communities should be at the forefront of business goals.

In the statement, the CEO of J.P. Morgan Chase & Co and chairman of Business Roundtable, made it clear:

> Major employers are investing in their workers and communities because they know it is the only way to be successful over the long term. These modernized principles reflect the business community's unwavering commitment to continue to push for an economy that serves all Americans.

This was followed up in November 2019 by a report called The Principles for Purposeful Business[15] published by the British Academy. The author, Professor Mayer, states that global crises such as the environment and growing inequality are forcing a reassessment of what business is for: 'The corporation has failed to deliver benefit beyond shareholders, to its stakeholders and its wider community. At the moment, how we conceptualise business is, it's there to make money. But instead, we should think about it as an incredibly powerful tool for solving our problems in the world'.

The advent of purpose-driven business is also illustrated by Mars chairman Stephen Badger admitting things are different now: 'We've never felt the need to be public but times have changed', he said. 'The talent [employees] really want to know what the company they work for, stands for. Equally important is that the magnitude of the challenges facing the world – climate change, poverty, biodiversity loss – these are issues that we care deeply about'.[16]

So the focus on size and power that characterised construction marketing and communications in the 1980s, 90s and early 2000s is being replaced with targets and aspirations to create a social legacy. The narrative now is about the number of lives improved and how companies are tackling the threats posed to society by reducing carbon and waste intensity, as well as recycling and investing in ways to offset what carbon is produced.

Customers want more

Customers are changing too and want more social value. Take the public sector. According to the Office for National Statistics, the value of new construction work in Great Britain in 2018 reached its highest level on record at £113,127 million, driven by growth in public sector work of £2,697 million.[17]

Political drivers and evolving priorities among public bodies now make purpose-driven companies with clear strategies on social value essential to who they choose to deliver their capital projects, supported by the 2012 Social Value Act which put on the statute book the importance of using public sector procurement to drive wider regional benefits like job opportunities, skills growth and sustaining local business.

It's not just about an end product. Construction must show the value of such work in boosting jobs, getting young people into work, supporting local companies and creating a legacy of skills.

The future of public sector procurement is about working with like-minded contractors that deliver this for their communities in a myriad of ways. This includes taking homeless people off the streets, equipping vulnerable young people with skills and self-belief to find work, sustaining jobs by employing local companies and volunteering people's time and skills to improve facilities that would otherwise not be possible.

Private and public sector customers want to work with progressive companies that have clear aspirations for promoting local opportunity, strong commitments for protecting and nurturing the environment and tangible ways to leave a local legacy of growth.

Against this background, the task for the PR professional and construction communicator is huge.

They have to guide and shape the way their company, and its people, stand out in a busy market and have clear, distinctive brand values and identity.

A platform for this is having a track record of capability, stability and longevity that instils confidence. At the time of writing, construction has been buffeted by headwinds from Brexit, resulting in reduced orders, while corporate failings overshadow the huge positives that happen each day across the country with brilliant projects delivered and many lives improved.

The focus of today's communicator is about dispelling some ill-informed perceptions of repetitive, low skilled work to one that shows the reality: a high-tech, multi-faceted industry that people should aspire to be part of, with the ability to shape the society we live in.

Added value through communications

Today's construction communicator must be flexible, adaptable and able to react quickly to situations created by the rise of new media channels. Just being good at media relations will no longer cut it.

The varied nature of the PR role requires a good appreciation of how a broad spectrum of activities adds value. This includes harnessing the multitude of channels now being used to share information, influence and bypass the more traditional channels of the pre-internet age.

Engaging with audiences requires giving them something they don't already have. Thought leadership is key – something mentioned later in more detail – and personalised communications is another way to compete for attention and be memorable when people are time poor and boredom thresholds are low.

We live in an age of instant communication available on handheld devices, and that's only going to get quicker. It's going to change work patterns as people harness the technology to go agile; the desk-bound 9–5 job was reaching the end of the road even before the tsunami of flexible working practices created as a result of the COVID-19 pandemic.

Agile working is still in its infancy, especially on construction sites, but that is already changing radically. Hot desks, home offices, working on the move and

68 *Andrew Geldard*

with more flexible hours – communicators need to adapt their output to reflect this.

For example, it is now normal for people to consume bite-sized portions of knowledge on their smartphones, while business and social digital platforms sites like LinkedIn, Instagram and YouTube further allow people to be publishers of their own content. Are you harnessing these platforms in a coordinated way to create positive content? For example, LinkedIn apparently says the platform sees 172,800 new users every day and about 62 million new users every year.[18] That's a lot of information sharing within networks.

Corporates as DIY publishers

Corporate use of digital media channels to publish news, engage audiences and build brands is now mature. In 2018, data from UMass Dartmouth (part of the University of Massachusetts) and Market US showed that among the Fortune 500 of top US companies, 92% used LinkedIn as their leading social media platform of choice. However, it shared the lead position with Facebook (92%), while Twitter dropped to third place (79%) compared to the year previously.[19]

Is your use of digital channels like LinkedIn, Facebook and Twitter about announcing or 'pushing out' news rather than engaging? And when people engage with you, are you ready to respond to make the most impact?

Most people are time-poor and want fast, easily digestible information, so engaging them has to reflect this. Tone of voice is key too, and needs to convey the dynamic, values-focused nature of construction firms and an industry that is cool to work for. There is also huge benefit to be gained from galvanising brand ambassadors and working with them to define the right channels for reaching out to their own communities. This not only takes messages to new and wider audiences but also makes people feel part of a purpose-based business.

Utilising colleagues as effective brand ambassadors means you amplify messages through their channels. Don't rely on colleagues to naturally 'like' your output. Empower them to become brand ambassadors as a powerful way to turbocharge what you say, reach a bigger audience faster and have the credibility that a peer is 'spreading the word' rather than through a traditional company channel.

Box 4.2 Case study: creating effective construction brand ambassadors

A brand ambassador is an advocate of the brand they work for. They use their network of channels to promote company messages by sharing positive, authentic content that creates brand awareness within their networks that is also spread virally a wider audience.

Willmott Dixon rolled out its brand ambassador programme in 2019. This was done through a series of workshops and webinars and supported by a Brand Ambassador Yammer group. The objective of the programme was to harness the immense publishing power of people's digital media channels to create a positive advantage and wider societal engagement for what the company is about. It involved:

- Explaining what a brand ambassador is and how this is beneficial to both the company and the individual through building their own professional profile.
- Educating people on key Willmott Dixon messages and values and how to communicate these effectively.
- Explaining different audiences and our corporate communications channels so that staff are aware of the tools at their disposal (blogs, website, podcasts).
- Understanding the approach to project/framework communications, encouraging teams to set up project hashtags to give real-time updates from their projects.
- Helping staff set up their own professional social accounts and empowering use by teaching best practice and top tips – what are they interested in; what is their motivation? Employees share updates on their projects/frameworks through their own accounts in a more authentic way than through corporate channels.
- Encouraging people to share content that resonates with them – their posts should always be authentic.
- Setting up a Yammer group where the communications team will post relevant news updates, content and supporting information for ambassadors.

The objective isn't just external, it's internal too – encouraging teams to embrace internal channels, share updates with their colleagues and help to drive engagement within their teams.

The training also included:

- Creating a recorded webinar that can form part of all new starter's inductions, so they are aware of our approach when they join.
- Training the regional communications managers to deliver the workshops as well to create more flexibility for delivery.
- Adding to online training platform My Learning and to communications playlists.
- Raising awareness via Yammer and the intranet.

Internally, communication needs to both create and support a culture of an informed and empowered workforce. Externally, it needs to emphasise social and environmental responsibility and the positive legacy that the industry leaves behind.

Contractors need to be nimble and flexible to stay in step with changing audience priorities and consumption needs as well as with new and evolving media channels and changes within the industry itself. For example, as more modular construction and automation comes on board, communicating how people will be retained on sites will become vital to supporting their wellbeing as well as preventing any further outflow of skills from the industry. Within this context, the role of the communicator needs to be continually re-evaluated to establish the skill sets needed and how to provide them.

The power of harnessing the new-age media (I don't like using the phrase 'social media') is about coordination to drive consistency. Before posting anything online, consider how it will further the organisation's overarching objectives and be memorable. There's a lot of dullness out there, why should a time-poor audience give a company special treatment unless it is adding value to their world?

That means building content to connect with your audience. If the plan is to highlight how great the company is to work for, especially to younger people choosing a career, then one way to connect is to use role models to inspire people, for example, to apply for the graduate manager scheme. That could be via a vlog or blog showcasing a recent graduate manager; both these options are low cost to produce but effective in the views and engagement they generate because they are sincere, and they are not the same old corporate sales pitch.

Box 4.3 Case study: communicating careers at Willmott Dixon

To reach out to teenagers and young people choosing careers, the best way to explain what a career opportunity at Willmott Dixon looked like was through their media platforms. Knowing that involved research to understand their world and not what people in their 30s or 40s think it is.

To achieve this, a series of workshops were held with people between 15–17 years and 18–21 years to understand their perceptions of construction, what they were looking for in a future career and what media they used for this information.

This found that traditional channels did not carry much influence and the sort of ideas being considered, such as adopting a version of the TV quiz *Who wants to be a millionaire?* were outdated as none of the young people really knew what it was. A classic case of someone 20 years older trying to guess what would be engaging to a 17 year old!

While the perceptions of construction were as expected, the more the focus groups learnt about the reality (office jobs, solving complex problems, good financial rewards), the more engaged the young people became.

The feedback helped refine and concentrate the imagery and content used to show young people in non-site environments talking about their career opportunities.

> It was deemed that a digital media approach using YouTube and Instagram, where recent graduates filmed their average day and gave commentary about life in the construction industry, was an effective way to get engagement. Non-traditional construction roles in areas such as HR and IT were also highlighted to show the variety of roles available.
>
> This led to ten graduate managers using their own smartphones to record films which were then put on the company's social media channels as well as shown at careers fairs. The results led to over 2,500 views on the YouTube channel and were a big factor in the 2018 graduate management scheme, attracting over 4,000 applications. Also, given that many of the vloggers were women, the 2018 graduate intake saw a parity of genders for the first time ever.

Younger people, as a rule, don't engage with corporations on their social profiles – it's not cool – so making the content something that they can relate to is the difference between effective communications and irrelevant 'noise' that goes unnoticed.

Another way to highlight a specialism or capability is through a podcast. This is surprisingly effective as a means for someone to impart something really useful and is easy for the audience to consume. The platform is easy to host too, and what's even more important, it provides something that bit special – communications with a personality.

Expensive production companies or studios are not needed to record a podcast. The equipment is straightforward to buy and software is easily available to edit the interview into a good podcast. This supports the notion of the brand ambassador – podcasts are important items of content for people to share on their platforms as publishers.

Dealing with a platform overload

The proliferation of platforms makes for huge opportunities to amplify messages, whether blogs, podcasts, video or webinars. It's worth reviewing a few of the newer channels and their importance to achieving business success. The following is my opinion on what works best and how to get the most from them.

- **LinkedIn** – a really good source of sharing information. Ensure that brand ambassadors make full use of it by sharing company news as well as liking the company updates. In my experience, LinkedIn gets more engagement and interaction than most digital media platforms. In a business context, communication is between those with a common purpose, rather than those who just want to chat.
- **YouTube** – short sharp clips that relate to watchable content perceived as non-marketing seem to pick up the viewing numbers. But it only really works

when you use the power of other platforms to encourage footfall to your latest post.
- **Twitter** – do people spend all their life on Twitter? Assume not – content should be of real importance and relevance to the audience otherwise it will not be registered. Twitter can be effective in allowing people to engage with local projects (both positive and negative aspects), which makes the company seem more approachable.
- **Instagram** – still popular with the younger generation, Instagram is a great way to tell a business's stories and convey the culture, which is important in enabling people to see past the stereotypes of an industry.

Content has evolved

Twenty years ago, size mattered. No more: the focus is now how we strengthen society's wellbeing, deliver social cohesion, provide life-long opportunities and be strong contributors to environmental sustainability. Platforms and channels are just methods for conveying this. The way that communicators add value is by devising a unique message of purpose, aligned with what our wider communities want to contribute to and empowering them to feel part of making it happen.

What makes a successful communicator?

I am often asked what makes an effective PR professional in construction, compared to one with a title but little real authority. First, it's not about doing the bidding of your nearest director (although it is about supporting and responding – that's absolutely vital). You are there to advise and counsel, to think, devise and recommend, not just wait to do.

I like to think of being an 'ideas factory' as a key skill, working with people to shape and implement ways for communications to improve business success.

Here are my top tips for being effective at shaping your company or, as a PR advisor, providing wise counsel and insight.

- **Build up your internal networks** – good communicators cannot work in isolation; they need colleagues to support them with prompt and timely information.
- **Evidence the value of what you do** – you should devise a quarterly report showing real statistics of website traffic, views of blogs, most popular tweets and LinkedIn posts. This allows you to also pick up trends and see what works and what is not popular.
- **Present ideas, don't wait to be told them** – be bold and imaginative and aim to develop at least one idea each month that will inspire others and be of real value.
- **Don't outsource your thinking** – there are no short cuts when it comes to developing solutions to business challenges. You need to own the detail of what you are doing and not rely on someone else to have it instead.

- **You're the custodian of your company brand** – and that starts with your own brand, so ensure that your peers look up to you with respect and regard you as the expert in what you do.
- **Don't let yourself be compartmentalised** – develop your skills and spot ways you can add value in non-core areas, even if that means taking a lead in strategic relationships that enter the field of business development. Don't think 'that's not my job', as in the twenty-first century it probably is, even if the company does not always recognise it at first.
- **Own the communications** – every week there are new opportunities to reinforce what your company stands for, to highlight a key culture characteristic and a reason why people should be proud to come to work. There's no one else who can better use the content and channels create the work environment that makes your company successful.

Notes

1 Construction activity crashes to record low, Construction News, 6 May 2020: https://www.constructionnews.co.uk/data/data-news/construction-activity-crashes-to-record-low-06-05-2020/ [Accessed 8 August 2020].
2 CIPS, Fastest rise in construction output since October 2015, 6 August 2020: https://www.cips.org/who-we-are/news/fastest-rise-in-construction-output-since-october-2015/ [Accessed 8 August 2020].
3 Construction Leadership Council, Roadmap to Recovery, June 2020: https://www.constructionleadershipcouncil.co.uk/wp-content/uploads/2020/06/CLC-Roadmap-to-Recovery-01.06.20.pdf [Accessed 8 August 2020].
4 10 Downing Street press release, 'Build, build, build': Prime Minister announces New Deal for Britain, 30 June 2020: https://www.gov.uk/government/news/build-build-build-prime-minister-announces-new-deal-for-britain [Accessed 8 August 2020].
5 CITB, Construction Skills Network forecasts 2019–2023, 25 February 2019: https://www.citb.co.uk/about-citb/construction-industry-research-reports/search-our-construction-industry-research-reports/forecasts/csn-forecasts-2019-2023-uk/ [Accessed 8 August 2020].
6 UK Construction Week, 'Where are all the women on the tools?', 7 March 2019: https://www.ukconstructionweek.com/news/where-are-all-the-women-on-the-tools--construction-buzz-207 [Accessed 8 August 2020].
7 McKinsey & Company, 'Delivering through diversity', 18 January 2018: https://www.mckinsey.com/business-functions/organization/our-insights/delivering-through-diversity [Accessed 8 August 2020].
8 https://www.youtube.com/watch?v=lpWW5XT6_2Q&t=46s [Accessed 24 January 2020].
9 Financial Times, 'Striving for inclusion: top European companies ranked', 20 November 2019: https://www.ft.com/content/bd1b4158-09a7-11ea-bb52-34c8d9dc6d84 [Accessed 8 August 2020].
10 Wates Diversity & Inclusion Plan, July 2019: https://www.wates.co.uk/wp-content/uploads/2019/08/Wates-Inclusion-plan-1.pdf [Accessed 8 August 2020].
11 Wikipedia on Carillion: https://en.wikipedia.org/wiki/Carillion [Accessed 8 August 2020].
12 Building, 'Balfour Beatty says 80% of sites were open last month', 1 June 2020: https://www.building.co.uk/news/balfour-beatty-says-80-of-sites-were-open-last-month/5106257.article [Accessed 8 August 2020].

13 Willmott Dixon, Our People: https://www.willmottdixon.co.uk/sustainable-development-reviews/2019-sustainable-development-review/our-2018-performance/social/our-people [Accessed 8 August 2020].
14 Business Roundtable redefines the purpose of a corporation to promote 'an economy that serves all Americans', 19 August 2019: https://www.businessroundtable.org/business-roundtable-redefines-the-purpose-of-a-corporation-to-promote-an-economy-that-serves-all-americans [Accessed 8 August 2020].
15 British Academy, 'The Principles for Purposeful Business', November 2019: https://www.thebritishacademy.ac.uk/publications/future-of-the-corporation-principles-for-purposeful-business?from=homepage [Accessed 8 August 2020].
16 BBC, UK 'has particularly extreme form of capitalism', 27 November 2019: https://www.bbc.co.uk/news/business-50562518 [Accessed 8 August 2020].
17 ONS Construction Statistics, Great Britain 2018: https://www.ons.gov.uk/businessindustryandtrade/constructionindustry/articles/constructionstatistics/2018 [Accessed 8 August 2020].
18 99 Firms, LinkedIn statistics: https://99firms.com/blog/linkedin-statistics/ [Accessed 8 August 2020].
19 UMass Dartmouth Center for Marketing Research: https://www.umassd.edu/cmr/research/social-media-research/ [Accessed 8 August 2020].

5 Communications for specialist subcontractors
Demonstrating value in the specialist supply chain

Cathy Barlow

Introduction

Specialist subcontractors face challenges and opportunities that are peculiar to their sector.

On the one hand, their services could be in high demand because there is a limited number of companies or individuals who can do what they do. On the other hand, if they are very niche, projects may be few and far between.

Competition can be intense, with just a handful of the same, equally skilled businesses pitching for the same work each time, and it may be difficult to persuade clients to change allegiance from subcontractors they have worked with previously. Competing on price is a dangerous gambit that is unlikely to end well – specialist services usually command a premium price, so a different approach is needed.

In the world of construction, 'value engineering' has now become a toxic term meaning 'opting for the cheapest solution'. Specialist subcontractors should be seen as part of value engineering in its truest sense, delivering value over time by making sure that buildings are constructed correctly in the first instance and continue to perform as they should. Therefore, they need to demonstrate why their more costly services deliver real value over the long term. The job of PR professionals is to help them to communicate this effectively, as well as to reach and engage with their target audiences.

Why do we need specialist subcontractors?

There are many different factors influencing the evolution of the construction industry today. The climate crisis has increased the pressure to build more energy efficient buildings, to consider the need for flood resilience and to meet the structural requirements that enable buildings to withstand more frequent extreme weather conditions.

Meanwhile, growing populations and urbanisation call for buildings that can be constructed quickly and that optimise space. Imperatives such as the preservation of life and property require an understanding of how a bewildering array of different products, systems and designs will react and perform in the event of fire.

Each of these areas has also led to an unprecedented growth in innovation to help meet market demand. We have seen the development of new, higher performing materials and an increase in the use of modern methods of construction, including modular buildings which arrive on site fully or partially assembled, off-site engineering and even 3D printed houses.

Digital tools such as Building Information Modelling (BIM) and Climate-Based Daylight Modelling (CBDM) are revolutionising the way that the construction process is carried out, from initial concept to detailed design and construction, right through to how a building is maintained and managed throughout its life cycle.

Meeting the different demands of building regulations, new types of product and highly technical processes often requires expert input and application. Specialist subcontractors are needed to make sure that buildings perform as intended, that there are no clashes between different elements of a design or construction details and that the aspect of the construction that they are responsible for is fit for purpose.

Examples of specialist subcontractors who might work on a project include (but are not limited to):

- Façade engineers
- Dry-lining specialists
- Curtain wall specialists
- Photo Voltaic (PV) installers
- Green roof installers
- Water management consultants
- Mechanical and Engineering (M&E) consultants
- Heating, Ventilation and Air Conditioning (HVAC) installers
- Audio Visual (AV) installers
- Architectural archaeologists
- Structural engineers
- Fire safety engineers
- Environmental consultants
- Interior designers
- Quantity surveyors
- Project managers
- Smart building control specialists

Some of these roles could be considered mainstream, particularly on larger projects, but they may also have specialists within that profession to deal with areas such as heritage sites or specific design requirements, such as Passivhaus.[1]

The range of specialists is clearly large, and the nature of the work they do can be completely different, but the guiding principles of PR and marketing that work for each of them remain broadly the same.

Meeting the challenges and finding opportunities

Helping a client to differentiate themselves is a key aspect of a PR consultant's work.

To do this effectively, it is important first to understand the direction the client wishes to take (for example, are they targeting a specific sector, such as local authorities? Where are the growth areas in the market?), what they want to achieve (brand awareness, expert positioning, improved customer experience) and where they sit in relation to the competition (market share, service offer, price point).

A bit of digging at the outset will lead to much better results overall and will also help to focus the client on what really matters to them.

Box 5.1 Four questions to ask

What are the organisation's business objectives?
What are the client's current strengths and weaknesses?
What are the priorities in the short, medium and long term?
What resources does the client have available?

Objectives

As already highlighted, construction is a complex industry with many influencing factors, so while a primary objective is often ultimately to grow sales, there may be many other aspects that need to be addressed in order to achieve this. As touched on above, this could include influencing regulation, impressing investors, recruitment and retention of staff, engaging with suppliers or educating bodies such as building control officers or local authorities. To clearly identify the objectives, therefore, it is crucial to understand the context behind the ultimate goal, and one of the most effective and straightforward ways of doing this is to do a SWOT analysis.

Strengths and weaknesses

Conducting a simple SWOT analysis can provide a useful insight into where the focus really needs to be. Note: it is not always where the client thinks it is!

In Table 5.1, a clear PR and communications priority should be to improve the content on the website. Another might be to provide educational content aimed at those responsible for commissioning dry lining subcontractors (such as developers, local authorities, facilities managers, building owners and private landlords). If they can be made to understand the consequences of using unskilled

Table 5.1 Example SWOT for a dry lining contractor

Strengths	Weaknesses
• Specialist service • Technical expertise • Good social media engagement • Professional accreditation • Work guaranteed	• Price point compared to unskilled workers • Lack of awareness in the market of the need for expertise • Poor website content
Opportunities	Threats
• Energy Performance of Buildings Directive (EPBD) driving refurbishment • Demand for improved housing standards • Minimum Energy Efficiency Standards (MEES) for rented properties	• Competitive market • Unskilled workers undercutting skilled tradespeople • Reduced access to skilled workers following Brexit • New methods of construction

labour (such as poor installations leading to inadequate insulation, damp, mould, structural damage, devaluation of property, remedial costs and impact on occupier health), they are more likely to appreciate the value of using a qualified professional.

One important point to note: a good SWOT analysis includes references to regulatory influences. A truly effective construction-focused PR campaign cannot be conducted without at least a working knowledge of this crucial aspect of the industry. It is important to get to know which regulations and standards affect the client's work and to be aware of any upcoming changes or consultations – this represents one of the best opportunities to showcase a client's expertise in their specialist area.

Priorities

The PR and communications priorities identified through a SWOT analysis will affect how the campaign develops and will also impact how easy it is to evaluate the effectiveness of the work.

Take the two priorities identified in Table 5.1. In the first instance, improving and building upon website content should be a relatively straightforward task. This is often fundamental to PR work, whether that is managing a website or simply providing the content. It is one of the easiest things to monitor and assess the impact of, provided access to website analytics through platforms such as Google Analytics is available. It is also an area where PR practitioners may be able to deliver real value as specialist subcontractors in their own right, as delivery of a properly optimised, useful website offering a good customer or user experience is often a weakness in the construction industry. In these situations, it is important to do a thorough analysis before work starts, so that there is something to benchmark against.

The second priority, that of educating the market, is a much longer game, and subsequently much harder to evaluate effectively. Some of the tools and tactics that can be used to produce the work will be examined later in this chapter, but again, it is important to benchmark at the start of the campaign if possible. This could be done through surveys, focus groups, analysis of competitors and market share, where enquiries are coming from and where bids have been lost.

Resources

All too often, PR and marketing agencies are asked to put together a proposal with no clear framework or budget. When a potential client asks what it would cost to run a campaign for them, the answer is 'how long is a piece of string?'. There is no point spending hours developing a great strategy that could achieve amazing things if there is no budget to support it. Getting at least a ballpark figure to work with is essential.

Bear in mind that the resources that can be used are not just reliant on budget, but also on what contacts, case studies, research, testimonials, photography, spokespeople, marketing and educational material the client already has. If there is no existing collateral, it is going to take longer and cost more.

It is also important to manage expectations at the outset. In the words of Tom Waits:

> Jim Jarmusch once told me Fast, Cheap, and Good... pick two. If it's fast and cheap it won't be good. If it's cheap and good, it won't be fast. If it's fast and good, it won't be cheap. Fast, cheap and good ... pick two words to live by.[2]

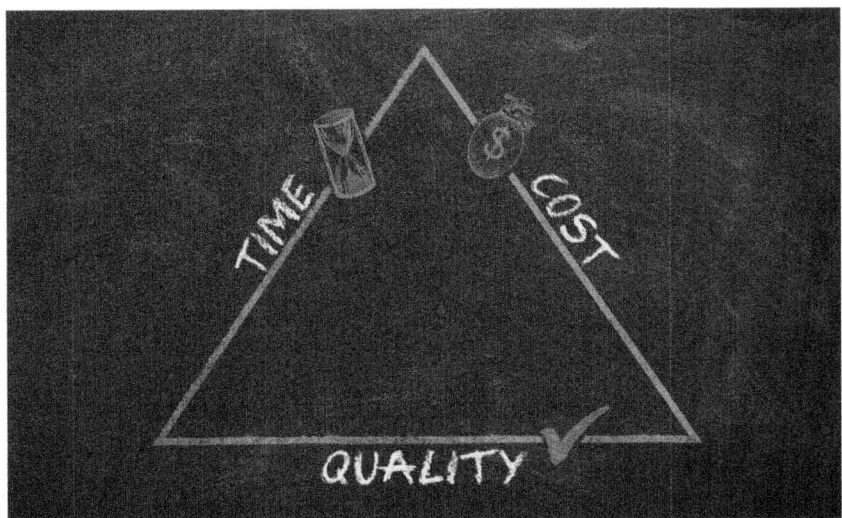

Figure 5.1 Time, cost and quality.

Be very clear with clients about the balance of time against quality and cost, to avoid being left juggling an impossible task. This is especially true of smaller businesses which may not understand that the value of PR in construction, much like their own value, often lies in the long-term impact.

Target audience

One of the primary challenges in any construction PR campaign is identifying the target audience. It's a challenge because the supply chain is often extensive, and lead times on projects can be years. Therefore, much depends on where the specialist contractor sits in the supply chain and timeline and what they are trying to achieve. For example, is the campaign about helping to drive the current pipeline of work, or is it about reaching influencers whose decisions will affect the building regulations or the pipeline of work in the future? It may be that the client wishes to build relationships with suppliers or to attract and recruit skilled workers – a difficult task in an industry that has a notorious skills shortage. Often, there will be more than one aspect to the work.

For communications to be effective, it is crucial to understand what matters to the target audience. What problems can the client help them to solve? How can they make things better or easier? The aim is to pull customers towards clients by offering them something that they want, rather than pushing things at them that they may not be ready for. Build relationships. That's what PR is about.

The next consideration is how the target audience finds and absorbs information. What platforms are they using? How do they engage with suppliers?

Box 5.2 Three questions to ask

Who is the target audience?
What do they need to know?
Where are they looking?

What to talk about

Once the target audience has been identified, the client's messaging needs to be positioned to appeal to that audience. There are just three things that motivate people to act – fear, desire and necessity. Tapping into one or more of these three motives in an appropriate way will make communications more powerful and meaningful to the people receiving them.

For example, it's no secret that the building regulations are dull, complex and demanding, but they must be complied with. Highlighting any upcoming changes and explaining how they affect customers (necessity), the consequences of not complying (fear) and how the changes might be turned to advantage (desire) could potentially hit all three triggers, providing useful information and a solution that pulls business towards the client.

Innovation can be another great draw, feeding into customers' desire for the latest shiny new toys and developments. However, before clients get carried away by their novel offering, remember that, in the context of construction, innovation also needs to be backed up by some serious research and proof that it works. This is a very traditional industry, and it can take a long time for new approaches to be accepted. The stakes are high, and often few people are willing to take a risk.

Thought leadership

In terms of finding something to say that might resonate with potential customers and stakeholders, there is no shortage of material. Once the focus moves from purely what the client offers and puts it into the context of the wider industry, it will immediately become more relevant and interesting to the target audiences.

Topics could include:

- The changing regulatory landscape
- Government initiatives
- Tests and standards
- New developments
- The circular economy
- Sustainability
- Social value (very important for public sector work)
- Aspirational standards such as BREEAM,[3] LEED,[4] Passivhaus and WELL[5]
- Regional developments (Northern Powerhouse,[6] Midlands Engine,[7] HS2[8])
- Fire performance
- Energy efficiency
- Waste
- The skills shortage
- Health and safety

The list goes on and on… Of course, not all of these may be relevant to the specialist subcontractor in question, but there will always be something in the wider topics to tap into.

Specialist subcontractors are experts in their field. They can expect to charge a premium fee for their services, and it is crucial that they can demonstrate why they provide value. 'Thought leadership'-type content offers the opportunity to do this, by pointing out the pitfalls of not using qualified professionals and highlighting the benefits of having a job done properly. From compliance issues to long term cost savings, there is a case to be made for using specialist subcontractors, and there are several ways to get the message across.

Technical articles and features

In the construction industry, print most certainly is not dead. Research from the Construction Media Index[9] confirms that architects in particular still like to access articles and information in print, with the majority apparently regularly reading hard copy journals. With literally hundreds of trade magazines covering construction and new publications appearing each year, print and digital trade media offers a rich source of communication.

Feature articles provide the ideal platform to deliver some really meaty or hard-hitting content that educates, informs and showcases the client's expertise. These can range from 700 to 2,000 words and can provide a basis for other work, including social media soundbites, blog content, presentations and CPDs (which are especially attractive to architects, as they have to attain 60 CPD points over the year if they are to maintain RIBA accredited status).

It is essential that articles are properly researched and supported by evidence wherever possible, not based on hearsay or speculation and not focused on 'competitor bashing' – such an approach is unhelpful and is not welcomed by the editors of quality publications.

Note: when pitching for an article in a publication, make sure the topic is relevant to the readership, and tie it in with the publication's own features list wherever possible. If in doubt, call the editors and ask.

Blogs and LinkedIn

Self-published material such as blogs and long-form LinkedIn posts are a valuable way of introducing key topics and of demonstrating the expertise of both the company and individuals within the company.

Note: if an individual employee is being used to inform and by-line blog posts, make sure the material will still be accessible if that individual leaves the company.

Events

Having a presence at events, particularly those with speaking slots, is a great way of showcasing expertise and solutions for common problems. Before the COVID-19 pandemic, there were lots of trade events up and down the country all year round, although the majority fell either in the spring or the autumn. They ranged from niche regional or sector specific events, such as EcoShowcase or the Built Environment Network events, to huge general construction shows such as Futurebuild or UK Construction Week. Speaking slots are usually tied into some kind of financial commitment, either sponsorship or being an exhibitor. We will need to see whether this events market recovers from the impact of the coronavirus disruption in 2021.

Of course, in the meantime, a lot of opportunities are opening for online events, and there is still a strong demand for expert speakers. Clearly, not every client will have the experience, resources or confidence to make a good speaker, however skilled they are at their job. It is important, therefore, to identify the right spokesperson and ensure that they are properly equipped to do a presentation, including training where necessary.

Show them how it's done

People need to understand what it is that the subcontractor does. They need to see the work in action and the kind of results the client can achieve. This helps potential customers to feel confident to engage, to be inspired or simply to understand that here is a solution to a problem they may have.

One of the most effective pieces of content is a good case study. It can be used in lots of different ways, across multiple platforms, and offers an opportunity to build relationships with third parties as well as clearly demonstrating the client's expertise.

It might be tempting to just crack on and write any case studies the client provides details for; however, it is important both to get the process right and to make sure that the case study is of a high enough quality and will be of interest to the identified target audience.

Good case studies can then be sent out as press releases to the trade press or as the basis for a feature length article illustrating how an issue has been dealt with. They can be used to provide good web or social media content or be turned into an award entry (the construction industry has many awards schemes). A video diary or time lapse photography can map progress, creating interest even before the project is complete.

Note: for refurbishment projects, make sure to get 'before' images as well as pictures of the completed work.

There are many different people and companies involved in constructing a building, and it is good practice to get the key players on board when developing a case study. This usually involves the building owner, architect and main contractor as a minimum. Asking permission to do the case study and seeking approval of the copy before distribution or publication provides the opportunity to make sure details are correct and to get vital testimonials and quotes. It also helps to protect the client (and the PR professional) if any of those details are wrong, as the third parties will have signed off on the copy. The process can take longer than the client might like, but it is well worth it to avoid any future conflict.

Note: make sure that everything is done over e-mail, or that telephone conversations are reflected back to the third party in an e-mail, so that there is an audit trail of what has been said and agreed.

As well as approving copy, permission can also be sought for taking photography of the installation or the finished project. The main contractor or architect may also have photography they are willing to share, which can help to keep costs down where budget is an issue. Architects are often happy to provide technical drawings, which can demonstrate how a specialist contractor has contributed

84 Cathy Barlow

towards delivering the project. Always ensure the right credits for copyright are given for imagery.

A picture is worth 1,000 words

Writers may not entirely agree with this saying, but it is certainly true that images help to bring a project to life and result in much higher levels of engagement, whether it is on social media, the website or getting picked up by an editor for print. A really good photograph can get the client on the front page of a key trade publication free of charge. It is also a great way to help people to visualise what their project could look like with the client's services.

Of course, photography is not the only kind of visual content that can demonstrate the value the client offers. Video is a powerful and popular tool to attract attention, and animation is a useful way to explain a complicated process.

Infographics are also handy for conveying otherwise dull material, such as routes to compliance or the potential benefits of different methods of construction, in an impactful way. They can also be used to raise important topics such as mental and physical health, which people are becoming increasingly aware of as an issue in the construction industry.

Awards

From high-profile general construction industry awards (such as the Building Awards), to discipline, sector or material specific awards (such as the CIBSE, Housing or Brick Awards), every year there are opportunities to showcase what

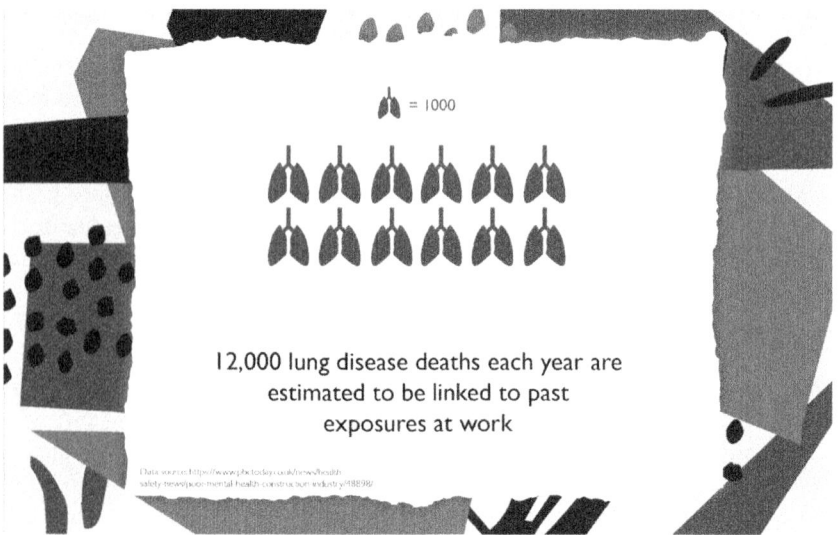

Figure 5.2 Using infographic-style design to help communicate key facts.

a client is capable of and to get their name in front of their prospective customers. Working collaboratively on an award entry can also help to strengthen those crucial third party relationships.

If a client has brought an innovative product, service or solution to the market, it might be worth exploring whether it would be eligible for a prestigious industry award such as the Queen's Award for Innovation, which carries a lot of weight and is valid for five years after winning. Be warned, this is not just about creating good award entry copy – it also needs to be supported with fairly detailed financial information too, so check first that the client is willing and able to supply that.

Website content

One of the fundamental mistakes many specialist subcontractors make is not explaining on their website homepage exactly what it is they do. The best-looking website in the world is useless if visitors cannot easily find the information that they need or even find the website itself when they are searching for a specialist service.

From getting basics such as clear signposting right, to making sure that content is fresh and relevant, a good PR professional can turn a website into the powerful tool it is supposed to be.

The kind of content that will help to do this includes:

- Case studies
- Blogs (thought leadership/knowledge base)
- Video content
- A clear indication of the services offered (backed up by case studies)
- Geographic reach

Before starting work, it is a good idea to do a full audit of the current content on the website, looking at how well it is optimised for search and how easily users can access the information they need and identifying areas that can be improved or built on. Tools such as Google Search Console are great for identifying successful keywords and areas that are weak. This is work that needs to be done regularly, as the market changes and people search for different information. It is worth noting that optimisation is an area that the construction industry is often poor at, which presents a good opportunity to boost the client's visibility over their competitors.

Note: don't forget to make sure that the branding, tone of voice and business persona/ethos is consistent throughout the website and is reflected in other material such as literature, social media etc.

Trade media

There is a clear hierarchy in the trade media, and different types of publications meet different needs in the market. For example, in the architectural press, there are the 'top tier' publications, which include titles such as *RIBA J* (the magazine of the Royal Institute of British Architects), *The Architects' Journal* (AJ), *Architecture Today* and the *Architectural Review*.

For a specialist subcontractor to get a mention in these publications, they would need to be involved in a featured case study, have some exceptional story to tell that architects would be deeply interested in or else pay for content. Editorial is created in-house by the publication, with the occasional opportunity to contribute an industry comment. The best way to build a relationship with these publications is to make sure to only share content that is truly useful or innovative, exclusive and relevant to their readers.

Then there are 'mid-tier' publications – magazines that will accept owned material, and which are a great vehicle for showcasing expertise. A story's potential for inclusion is greater if supported through advertising, but many editors will take articles purely on merit. If advertising and editorial are both being booked in, make sure they appear in different issues of the magazine, preferably editorial first. That ensures the client gets their name in front of their target audience in two issues of the magazine instead of one, and that the editorial does not appear to be paid for, lending it greater authority.

Other types of publications include 'product books', which are essentially used as product reference tools for specifiers and designers searching for new products and applications. Case studies can work well in these too.

Finally, there are the specialist publications for areas such as HVAC (heating, ventilation and air conditioning), modern methods of construction, steel construction or Passivhaus and the official magazines for trade associations, such as *CIBSE Journal*, *Construction Manager*, *The Structural Engineer* or *Modus*. This last category is particularly useful for targeted thought leadership pieces.

One thing to be aware of with some parts of the specialist trade press is the weird phenomenon of 'colour separation' charges. This hangover from the days of dealing with physical photographs is one way that the trade press has mitigated loss of income from dwindling advertising spend. Clearly, it counts as paid-for coverage, but it is relatively inexpensive and may help to build relationships with the publications, opening the door for more substantial pieces of owned editorial.

Getting social

These days, no PR campaign would be complete without integrating social media into the communications strategy. As a sector, the construction industry was a late adopter of the medium, and many businesses still struggle to do it well, if at all. This may be a result of lack of resources and time, but more likely it is a lack of understanding in how to use the different platforms effectively.

It is not simply a question of understanding which platforms the target audiences are likely to be using; it is also a question of changing the messaging to suit the platform rather than pushing the same content out over everything.

For example, architects are probably the most adept and earliest adopters of social media in the industry; they will often use Twitter for comment, information and breaking news and Pinterest and Instagram for inspiration, sourcing suppliers and sharing projects.

Facebook has become more widely used for business-to-business (B2B) communications, but is still generally considered a consumer-facing platform and a very effective tool for reaching that market.

LinkedIn is still mainly used for growing networks and is a great platform for thought leadership pieces and engaging in debate.

Video is both popular and powerful, whether it's a quick vox pop on Facebook Live, a 'how to' video on YouTube or a fully produced educational piece hosted on Vimeo.

Don't forget the growing popularity of the podcast, which is fast becoming one of the most important platforms for targeting the upcoming generation of construction professionals.

Research from Rajar[10] released in Spring 2018 showed that 6 million adults in the UK listened to a podcast each week, compared to just 3.8 million in 2016. The demographic is important, with research from Ofcom[11] showing that 49% of podcast listeners are under 35. Although there will always be a place for print and visual platforms, this is where there may be a serious shift in how people prefer to absorb information, as the younger generation start to become the decision makers in businesses.

Social media is an ever-evolving beast, with new platforms constantly appearing. Stay focused on those that the target audiences are known to engage with, but be aware that trends will shift, especially with younger demographics, so stay up to date with any developments and monitor who is on them. According to inbound marketing experts Hubspot,[12] the best place to keep up with social media trends is, of course, on social media. Using Twitter hashtags, Google Alerts, subscribing to relevant blogs and attending social media conferences are just a few of the ways to stay in tune.

As already mentioned, PR is all about helping to build relationships – with customers, the media, suppliers, influencers, specifiers, investors, employees, third parties and any one of the many stakeholders involved in the complex world of construction. We do this by providing clear communications, encouraging dialogue and engaging people across multiple platforms.

Apart from the tools we have already explored, there are other ways to help build those relationships.

The customer journey

Talking to customers might seem like an obvious thing to do, but it's surprising how many people don't give much thought to the customer experience. There are several clear touchpoints on the customer journey that provide an opportunity to build relationships and find out what customers really want.

> **Box 5.3 Four questions to ask**
>
> How easy is it to navigate the website and find information?
> How easy is it for customers to get in touch, and are enquiries dealt with promptly?
> Is there a way to check customer satisfaction at the end of a project?
> Is there a way of staying in touch to help drive future projects?

From a PR point of view, how to improve the website experience has already been mentioned. Making sure that contact forms and details are clearly displayed and easily accessible is an important part of that process. To check on customer satisfaction, a quick survey from the client may be useful or a personal call to check that everything is ok. This is also an opportunity to get testimonials or ask people for ratings on platforms such as Feefo[13] or Google, which will help to boost visibility.

One way of staying in touch is by inviting customers to sign up to a quarterly newsletter that provides useful information and industry insights. To remain GDPR compliant, make sure they have clearly opted in, and always include an unsubscribe option on any communications of this nature.

Networking

A lot of people struggle with networking. It can be daunting walking into a room full of strangers and just starting to talk to them. However, meeting people face-to-face is one of the best ways of building a long-term relationship, so encouraging the client to attend these events is often worthwhile. PR professionals can potentially provide support by going along as well or, better still, providing some guidance on how to handle networking effectively.

> **Box 5.4 How to network in a nutshell**
>
> | Do | Don't |
> |---|---|
> | Find out what the person you are talking to does and where their interests lie. | Try and sell yourself. |
> | Have a knowledge of the latest industry developments – sharing information and expressing an opinion helps to showcase expertise and impress potential clients. | Form a closed group – you want to invite others to join the conversation. |
> | Take some business cards! | Dismiss people who don't seem directly relevant – you never know who their contacts are or whether they might become relevant in the future. |
> | Also, remember to get a card or make a note of the contact details of the person you've been speaking to. | |

Opportunities to network might come from events, trade shows and, of course, virtually on LinkedIn.

Crisis communications

How a client reacts in a crisis is one of the most important aspects of preserving those precious stakeholder relationships. It may not be something the client has considered because they do not expect things to go wrong, but it is crucial to be prepared, and a PR professional is often the first line of defence.

In the construction industry particularly, the consequences of something going wrong can be very serious indeed. Having a plan in place, with agreed processes and responses to potential crisis situations is a useful, if not essential part of any holistic PR strategy.

Measurement and evaluation

Because the true impacts of a PR campaign may not be realised for months or even years in construction, proper evaluation can be extremely difficult, if not impossible to achieve. However, certain aspects can certainly be measured.

Print

The obvious place to start is by logging the coverage achieved, both paid-for and free of charge. Print publications usually supply circulation/readership figures, sometimes with a breakdown of sectors subscribers are from. Look for ABC certification of the circulation figures where possible. Remember that the actual reach may be far greater, as publications get passed around practices. A small number of magazines list enquiry numbers on each piece and will forward details if they receive any responses. It is also worth monitoring goal conversions and events in Google Analytics which can help track spikes in traffic, related web page visits, increases in followers, sales enquiries or other goal conversions directly linked to the interest generated by online editorial.

Make sure to differentiate between paid and earned coverage. Advertising and coverage arising from colour separations fall into the first category, and coverage about the client generated by journalists or editors of the publication into the last. This last category is considered the most valuable, as it has the credibility of coming from an independent third party, provided, of course, that the coverage is positive.

Online

Many print magazines which also publish online can supply page view data. Some may also be able to provide link clicks, shares, enquiries and comments. You could also use a specific landing page URL or link tracking to make it easier to track performance via your web analytics and referral traffic.

Digital

If booking a digital package such as a newsletter or solus e-mail, publications should be able to supply rich data, including subscriber statistics, email opens, link clicks, shares and comments. Again, using a specific landing page URL, link tracking or newsletter sign up link will make it easier to track performance.

The impact of social media content is probably the most straightforward to track. This can be done by looking at any increase in followers, shares, comments and engagement. When receiving email enquiries, it is useful to ask where people found the information. It's also possible to track mentions via alerts (e.g. Warble Alerts for social and Google Alerts for the internet).

A labour of love

If specialist subcontractors face a lot of challenges, so do PR professionals. Operating as a PR professional in the construction industry is demanding, rewarding and downright hard work, but it is precisely these challenges that make it such an exciting sector to get involved in. The potential is huge for the profession to make a real difference, not just to clients and their business, but by contributing well-researched, useful and informative content that will help to guide those involved in creating the buildings where people live, work, learn and play. Specialist subcontractors need specialist support, and PR professionals are the right people to supply it.

Notes

1 The Passivhaus Trust – www.passivhaustrust.org.uk/ [Accessed 29 September 2019].
2 Tom Waits website, 20 May 2008: http://www.tomwaits.com/press/read/15/TOM_WAITS_TRUE_CONFESSIONS/ [Accessed 8 August 2020].
3 BREEAM – www.breeam.com/ [Accessed 29 September 2019].
4 US Green Building Council – new.usgbc.org/leed [Accessed 29 September 2019].
5 WELL – www.wellcertified.com/ [Accessed 29 September 2019].
6 Industrial Strategy: the Northern Powerhouse – northernpowerhouse.gov.uk/ [Accessed 29 September 2019].
7 The Midlands Engine – www.midlandsengine.org/ [Accessed 29 September 2019].
8 HS2 – www.hs2.org.uk/ [Accessed 29 September 2019].
9 Competitive Advantage, Construction Media Index research: https://cadvantage-knowledge.co.uk/product-tag/communication-strategy/?product_cat=report [Accessed 8 August 2020].
10 RAJAR Midas Audio Survey Spring 2018 – www.rajar.co.uk/docs/news/MIDAS_Spring_2018.pdf [Accessed 29 September 2019].
11 Ofcom – www.ofcom.org.uk/about-ofcom/latest/media/media-releases/2018/uk-podcast-listening-booms [Accessed 29 September 2019].
12 Hubspot – blog.hubspot.com/marketing/keep-up-with-latest-social-trends [Accessed 29 September 2019].
13 Feefo – www.feefo.com [Accessed 29 September 2019].

6 Communications for construction products

Promotional practices from specification to installation

Louise Morgan

Introduction

The construction sector is vast and, from a communications perspective, complex. Where product supply is concerned, manufacturers are faced with the challenge of keeping their brands front of mind at every stage of the build process, which is literally from specification to installation, and in some cases, through to service and maintenance.

Communications programmes therefore need to align with the flow of the supply chain, spanning architects and designers through to main contractors, specialist subcontractors and builders' merchants.

A changing landscape

While 'traditional' PR is generally not associated with direct lead generation, in the construction products sector it has historically played a key role in sales and marketing, making a quantifiable contribution to generating sales enquiries. In this respect, many building product PR campaigns fall much more into the 'paid' category of the PESO model.[1]

Only a decade ago, reader enquiries were a common feature in the construction press. Managed by the publications, readers were able to submit requests for further information in response to product news articles featured in the press. However, the rapid rise of digital communication has seen a huge decline in the need for this service, with potential specifiers or purchasers instead visiting company websites directly or making contact through email. Consequently, the role of PR in promoting construction products has changed.

The decline of reader enquiries has occurred simultaneously with a noticeable reduction in the breadth and depth of the construction trade press, with the number of construction media outlets now standing at 376.[2] This trend is not surprising given that all sectors have observed a decline in print advertising spend, leaving publishers with the challenge of producing quality editorial without the financial support of advertising.

That said, compared with other sectors, such as education, which has just 143 dedicated publications,[3] the construction industry still retains a vibrant and diverse media base, making it an effective communications tactic for product manufacturers.

Construction products and PR

Despite media relations activity no longer representing a reliable route for generating sales enquiries, there remain appreciable benefits to undertaking PR campaigns for construction product manufacturers.

The complexity of the construction supply chain requires companies to communicate with stakeholders at every stage of the build process.

A powerful communication channel, media relations activities can provide a breadth and depth of coverage that is often more cost effective than other tactics such as direct marketing. Moreover, it gives companies an opportunity to educate the market in a non-promotional way.

Construction product manufacturers will typically undertake a PR programme to achieve the following objectives:

- Launch new products
- Evidence product performance
- Align solutions with industry trends/regulation
- Raise brand profile/achieve thought leadership
- Create a warm sales environment

From a product supply perspective, the headline objective will always be to support sales activity and to drive commercial demand. In the event of an accident or incident in which a product is implicated, this would fall outside of the run rate PR activity and will not be addressed within this chapter.

The role of the builders' merchant

As addressed at the start of this chapter, manufacturers are faced with the challenge of keeping their brands front of mind at every stage of the build process, literally from specification to installation, and in some cases, through to service and maintenance.

However, in the context of the supply chain as outlined in Table 6.1, builders' merchants represent a critical point in the purchasing process.

As the main procurement contact for many subcontractors, it is the builders' merchants who have the most influence over product purchase as they are the interface at the point of sale.

Table 6.1 Key stakeholders at each stage of the build process

Build stage	Stakeholder
Planning	Developer/client
	Planning consultant
Design	Architect/designer
	Specialist consultants (e.g. fire, sustainability, quantity surveyors)
Construction – foundations	Civil engineer/groundworkers
	Builders' merchants
Construction – building envelope	Main contractor (quantity surveyor/contracts managers/site managers)
	Structural engineer
	Specialist subcontractors (timber frame, brickwork, façade, roofing)
	Builders' merchants
Construction – services	Specialist subcontractors (e.g. mechanical, electrical, renewables)
	Builders' merchants
Construction – fit out	Specialist subcontractors (e.g. flooring, dry lining, decoration, joinery)
	Builders' merchants
Completion	Building Control
	Main contractor (site manager, contracts/project manager)

Consequently, builders' merchants are a major target audience for many construction product manufacturers – particularly for commodity materials which tend not to be supported by a technical specification.

In a sales-driven communications strategy, the following tactics are effective in engaging with this audience:

- Coverage in builders' merchants press, e.g. *Professional Builders Merchant*, *Builders' Merchants News*, *Builders Merchant Journal*
- Trade counter literature – equipping the branches with 'cheat sheets' and materials which help them to better understand a product offering and where it sits in relation to competitive offerings
- Roadshows – visiting the merchant branches to engage directly with the customer-facing sales teams and encourage brand buy-in
- Industry events – including exhibiting at the National Merchant Buying Society Ltd (NMBS) and other buyer society events
- Competitions – monthly sales incentives which reward branches and individuals for reaching set targets

94 *Louise Morgan*

With builders' merchants, any communications programme must be closely aligned with the sales team, as forging and maintaining strong personal relationships is important with this audience. The role of communications should be supportive and help to create opportunity on scale, but overall campaign effectiveness will rely on trusted sales relationships.

Constructing PR campaigns

There is a range of standard tactics which are commonly used as part of a construction product PR campaign:

- Press releases
- Feature articles/bylined articles
- Industry awards
- Industry events (exhibiting/speaker opportunities)
- Case studies

Programmes are built using the most appropriate tactics for the product. For example, industry awards may not be suitable for a new product launch if evidence of the product in application is required as part of the submission criteria.

As regards media relations, budget should always be considered carefully during the planning stages, as there is often a 'pay to play' model in place for product news and articles. Some building product manufacturers will not participate in this sort of PR, but for others, it can represent a significant amount of their involvement with the specialist trade press.

Building a product-based campaign which focuses on issuing a press release per month to the target construction titles is unlikely to secure sufficient coverage without the support of 'placement fees'. Historically, these have been referred to as 'colour separation' fees (or 'colour seps') but in recent years have become more commonly known as 'PR panels'. These are, in effect, small advertisements/ advertorials, often based on product topics such as security, fire safety or accessibility or clustering a series of products' marketing messages in sections of a magazine dedicated to a specific sub-sector such as bathrooms, kitchens, lighting or roofing.

When to use a PR panel

While many businesses believe that publications which run PR panels are inferior editorial titles, the reality is that they may still provide a viable route to securing exposure when used in the right way:

- Only pursue PR panels with publications which can prove they reach the target audience and provide the 'pay to play' opportunity in response to the distribution of a press release.
- Avoid publications which speculatively target PR professionals with last minute offers of PR space at a vastly reduced rate.

Communications for construction products 95

- Budget-wise, a typical PR panel should cost in the region of £90–£120.
- Consider whether your news story can be communicated effectively in the allocated space, which in some cases will be under 100 words.
- Usually, PR panels are most suited to product launches or a case study summary.
- The use of a striking image is often essential in order to stand out on the page.

Bylined articles

In the context of product PR and promotion, one technique that is particularly effective with the specialist trade press is the placement of bylined articles. Unlike 'PR panels', these will usually be secured based on editorial merit and therefore do not incur a placement fee.

To successfully integrate bylined articles into a media relations campaign, the content must be educational and not promotional. Bylined articles will typically address a topical challenge in the market, such as a change in building regulations or advances in construction techniques where best practice advice can be shared.

Expos and online events

Before the COVID-19 pandemic put a halt to almost all mass events in 2020, industry trade shows represented an excellent opportunity to engage with the target audience face-to-face and, in many instances, to demonstrate products on-stand. Many of these opportunities are expected to open up again in 2021.

But in the meantime, online and smaller face-to-face events which operate a seminar or speaker programme are still a prime communications target. In the same way that bylined articles allow the client to present product messaging in a non-promotional context, presentation opportunities are a tried and tested tactic for educating an audience and reinforcing a brand's position as an expert in the marketplace.

Within the construction sector specifically, there is one main expo, the UK's largest built environment event – UK Construction Week – which spans the full industry, with dedicated zones, such as timber, for further targeting. In 2019, the show attracted more than 30,000 visitors, and it is expected to return in 2021.[4]

That said, as UK Construction Week has a very broad audience, many construction product manufacturers seek out more specialist trade shows which better align with the performance benefits of their offering. Examples of these include:

- Firex – ideal for the promotion of fire protection products
- IOA Conference (Institute of Acoustics) – appropriate for sound reduction-related solutions
- Concrete Show – suitable for foundation and concrete ancillaries

Case studies

The role of case studies in a construction product PR campaign is fundamental.

Stakeholders in the construction supply chain actively seek out examples of a product in application where its performance credentials have been verified on-site. This is particularly true where a modern method of construction (MMC) product is being brought to market, or where performance has previously only been verified in testing facilities as opposed to on-site.

While case studies are the backbone of many construction product PR campaigns (more so than in other B2B campaigns), sourcing and developing detailed case studies can be challenging.

The production of any case study which features a specific brand and supporting testimonial should always have full written sign-off from the client. This is best practice in any industry. However, this process becomes increasingly complex in construction product PR because of the supply chain complexity (as detailed in Table 6.1).

From a sales perspective, the construction product manufacturer strives to achieve a named specification at building design stage. However, when this progresses through tender stage to build, the quantity surveyor can seek to change the specification to an alternative, but similar product. Even if the specification remains intact, the product would not necessarily be purchased by the main contractor. Instead, the responsibility is generally devolved as a 'supply and fit' package to a specialist subcontractor.

At the point of handover, therefore, when a product has been installed and is performing as expected, the initial challenge is to identify which stakeholder is responsible for approving the creation and publication of a case study.

On larger, more high-profile projects associated with major brands, a clear PR protocol will usually be in place whereby every stakeholder is briefed regarding the production of case studies. However, this is not always the case, which can result in confusion as to who can provide the official permissions to go ahead with a case study.

Where a construction product has been installed by a specialist subcontractor, it is usual practice to seek permission from that installer first before pursuing a case study. Once that has been obtained, the most robust and substantial case studies which truly showcase the benefits of a construction product would cover:

- Architect/specifier perspective – why was the product specified? What challenges was it addressing for building regulations or compliance?
- Main contractor/client – addressing the wider objectives of the project and how a product/manufacturer support contributed to this.
- Subcontractor/installer – how easy the product was to work with on-site and detail on any technical support provided by the manufacturer.

As this demonstrates, developing quality case studies in the construction products sector can be very challenging. When planning PR programmes, it is important

Table 6.2 Approved documents (building regulations for England)

Part	Title	Contents
A	Structure	Building structure, structural loading, ground movement potential and disproportionate collapse
B	Fire safety	Warning and escape, internal fire spread for building lining and structure and external fire spread for domestic and non-domestic properties
C	Site preparation and resistance to contaminants and moisture	Site preparation including removal of contaminants and ensuring the site is resistant to moisture
D	Toxic substances	Ensuring fumes from foam-based cavity insulation do not enter the property
E	Resistance to the passage of sound	Protection from unwanted sound and reverberation both from within a home and from separate parts of buildings and adjoining buildings
F	Means of ventilation	Ventilation and airflow in domestic and non-domestic buildings
G	Sanitation, hot water safety and water efficiency	Supply of hot and cold water and provisions for washing facilities, bathrooms and kitchens
H	Drainage and waste disposal	Drainage of water, waste-water treatment, construction around sewers and storage of solid waste
J	Combustion appliances and fuel storage systems	Air supply, carbon monoxide warning, combustion product removal, storage of liquid fuel and protection against pollution
K	Protection from falling, collision and impact	Protection from falling, impact and collisions within a building
L	Conservation of fuel and power	Power and fuel conservation in new and existing dwellings and new and existing non-dwelling buildings
M	Access to and use of buildings	Accessing and using buildings, whether a dwelling or otherwise, including access to sanitary conveniences
N	Glazing safety	Ensuring sufficient glass and glazing around a property and that windows and ventilation points open and close safely and are accessible for cleaning
P	Electrical safety	Electric safety in dwellings
Q	Security in dwellings	Standards for doors and windows to resist physical attack by a burglar
R	Physical infrastructure for high speed electronic communications networks	Ensuring physical infrastructure for copper or fibre-optic cables and installation of wireless devices

to be realistic about what can be achieved with case studies. If approvals cannot be secured, consider producing lighter-weight content that profiles project details without referencing client names. In this context, the content is suitable for use in general sales materials but will not provide sufficient substance to support bylined articles or speaker presentations.

Putting product PR into practice

As with any media relations campaign, the most successful results are realised when activity is sustained over time. As a general guide, any new programme should be targeted to run for a minimum of six months.

As well as developing a clear understanding of the company and the construction products it manufactures, it is important to gain broader insight into the wider industry challenges and targets.

The UK Building Regulations[5] are a major driver of construction product specification, which makes them an excellent starting point for positioning PR messages.

Government initiatives, standards and targets are also a valuable reference for aligning product messaging, for example, the Government's 'Construction 2025' industrial strategy,[6] London emission targets[7] and BREEAM.[8]

Profile beyond product

While PR represents a productive communications tactic for helping to drive product sales in construction, for major brand players in the industry it is also a valuable strategic management discipline for:

- Corporate social responsibility
- Strategic positioning
- Thought leadership

In this context, the messaging will quite often move away from the technical performance of a product and focus more on a manufacturer's operations or industry credentials.

Trade associations and industry bodies are commonplace in construction, with manufacturers commonly using these as a vehicle for lobbying or education campaigns. For example, with the rise of the sustainable and zero carbon agenda, concrete is perceived as a less environmentally sustainable specification than timber, based on timber being a renewable material source.

However, the Mineral Products Association established a specialist division, The Concrete Centre, to 'provide material, design and construction guidance to those involved in the design, use and performance of concrete and masonry to realise the potential of these materials'. This includes guidance and publications which address the role that concrete has to play in areas such as energy efficient buildings and promotes the 'Concrete Industry Sustainable Construction Strategy'.

Communications for construction products 99

There will quite often be opposing sides of a debate in the context of which building materials are the best to use, but manufacturers can leverage the power of industry bodies such as these to ensure their opinions are heard collectively.

Sustainability is high on the agenda for construction product manufacturers, not only to demonstrate the lifecycle and recyclability of their own materials, but to also justify manufacturing operations. From a thought leadership perspective, this represents a strategic PR opportunity, as companies can use significant environmental changes and positive messaging to elevate themselves above the competition. As regards campaign planning, therefore, the sustainable commitment of the company is often prioritised above the product offering.

Box 6.1 Insulslab

In the UK, foundation methods typically fall outside of the Part L[9] compliance remit. This is because traditional systems, such as suspended slabs, beam and block floors or raft foundations, are classified in groundworks packages and do not form part of the exterior envelope. In response to the increasingly stringent thermal performance specified by the building regulations, a new system was developed which integrates high levels of insulation into the foundation structure.

As a potentially disruptive advancement in construction techniques, the foundation system Insulslab undertook a sustained PR campaign targeting:

- Housebuilders
- Ground workers
- Architects/specifiers

Media relations was identified as the most effective communication channel for:

- Reaching a diverse audience
- Communicating a complex message
- Educating the market on changes in construction methods
- Changing perceptions on the role of the foundation in building design

The key target press to reach the target audiences included:
Housebuilders

- *Professional Housebuilder and Property Developer*
- *Housebuilder*
- *Housing Specification*
- *Professional Builder*
- *Selfbuilder and Homemaker*

Groundworkers

- *Ground Engineering*
- *Construction News*

Architects / Specifiers

- *Building Engineer*
- *AJ Specification*
- *Architects Datafile*

While Insulslab provided a solution to address many challenges facing housebuilders and architects, the company recognised that the system would require groundworkers to adopt a new way of working. As such, PR activity was broken down into messaging targeted at two specific audiences:
Groundworkers

- Enhance competitive advantage
- Improve consistency on site
- Increase efficiency on site
- Explain how methods of foundation construction are becoming more innovative

Housebuilders/specifiers

- Improve overall U-value/thermal performance of building
- Reduce height of ground floor build-up while achieving superior insulation performance
- Increase speed of build and predictability of groundworks phase

Tactics

For the first 12 months, the campaign focused heavily on the placement of bylined articles with the main objective of changing perceptions and educating the market on how advances in foundation technology had addressed many challenges. Over time, as the system was adopted and used on site, the focus shifted to case studies to demonstrate Insulslab in application. The endorsement by major housebuilders built confidence in the market and encouraged further use by the groundworkers. Additional credibility was added by pursuing building industry awards which promoted innovation.

Results

Widespread and sustained coverage was secured in the target press, effectively educating the audience on changes in foundation techniques. Over time,

Insulslab became established as a standard foundation method. This required the PR campaign to change focus. In place of education, activity moved towards the production of technical case studies with the specific objective of supporting future sales activity. This demonstrates that the PR strategy of launching the system and educating the market was effectively achieved.

Client's comment

> At the point of launch to market, Insulslab was a genuine innovation in foundation techniques. We recognised that the breadth of coverage we needed to achieve throughout the construction supply chain would not have been feasible with our sales team alone. We were able to use the technical and by-lined articles to evidence our expertise in the sector and very quickly established ourselves as thought leaders.
>
> Liam Colebrook, New Product Development Director

Box 6.2 Trade Fabrication Systems (TFS)

TFS, a processor of coated and laminated wood-based panel products, made a significant £500k reinvestment into production facilities which further enhanced its surface treatments to include additional finishes such as wax and oil as well as CNC machining capabilities. This resulted in an increase in both capacity and capability which it sought to promote, the objective being to generate business from multiple revenue streams. This strategy was designed to reduce its current reliance on one large customer.

Having returned to the Timber Expo event[10] for the first time in several years, TFS put a stake in the ground from a marketing perspective. Howard Morris had taken over as managing director after serving ten years with the company, and in 2019, TFS celebrated 21 years of trading. These factors culminated in a positive PR opportunity on which the business aimed to capitalise.

As a B2B processor, TFS does not manufacture or directly ship its own products. Instead, the company processes value into existing substrates by applying materials such as coatings, oil and wax or pre-machining products such as doors. PR therefore presented an ideal tactic for educating end users on the potential benefits of off-site finishing as well as encouraging builders' merchants to expand their portfolios without increasing stock inventory.

The campaign targeted the following audiences:

- Housebuilders
- Builders' merchants
- Warehouse and logistics

Media relations was identified as the most effective communication channel for:

- Engaging with a diverse audience
- Communicating a complex message/educating the target audiences
- Promoting the benefits of off-site finishing and processing to end users and builders' merchants
- Showcasing case study examples to spark inspiration with other substrate manufacturers

The key target press to reach the target audiences included:

Housebuilders

- *Professional Housebuilder and Property Developer*
- *Housebuilder*
- *Housing Specification*
- *Professional Builder*
- *Selfbuilder and Homemaker*

Builders' Merchants

- *Builders' Merchants News*
- *Professional Builders Merchants*
- *Builders Merchants Journal*

Architects/Specifiers

- *Architects' Datafile*
- *Architecture Today*
- *Specification*

Flooring

- *Contract Flooring Journal*
- *Tomorrow's Flooring*

Timber

- *TTJ*
- *Structural Timber*

Key messages:

- TFS is completely independent – offering an agile and professional surface finishing service.

- TFS sells a service, not a product.
- TFS adds value to products without an inventory cost.
- Don't get lazy – create commercial advantage with creativity.
- Quality matters – TFS is certified to established industry standards such as ISO 9001.
- Complete design flexibility – we can match any RAL colour.[11]

Tactics

The media relations activity incorporated a mix of news announcements and proactively placed in-depth articles. Angles were:

News stories

- 21 years of trading
- Howard Morris's appointment as managing director
- Investment in production facilities
- Launch of a new range of finishes
- Project stories (door processing)
- Launch of training seminars for merchants
- Door processing service

Feature angles

- Building value into the supply chain – merchant-focused piece that educated readers on the value that can be realised with a surface finishing partner (without an inventory cost).
- 'It won't be all white' – article which focuses specifically on doors and why fit-out companies are missing an opportunity to provide access to a perceived premium finish.
- Bringing off-site 'in-site' – an examination of why the fit-out/flooring industry could benefit from an off-site approach with finishes.

Leveraging a series of innovative partnerships that TFS had developed with coating and timber manufacturers, the campaign also focused on the submission of award nominations.

Results

A steady stream of coverage aligned with the company's messaging was consistently secured, with article content repurposed for use on the TFS website and social media channels.

Client's comment

> While we provide a service that adds high levels of value to the manufacturers, merchants and end users, communicating this can be

> complex – as it's not a product as such. We felt a PR campaign that enabled us to consistently reinforce our messages in a more educational and less promotional way would help to prime the market and create a warm sales environment.
>
> Howard Morris, managing director, Trade Fabrication Systems

Editor's postscript

It is worth noting at the end of this chapter the changes likely to impact this area of construction communications in the near future. In particular, the promotion of building products is likely to be significantly influenced by the recommendations of the Hackitt Review[12] following the Grenfell Tower tragedy in 2017 and the activity of the Industry Safety Steering Group which is implementing those recommendations.[13]

Dame Judith Hackitt was highly critical of the way construction products are marketed and how product information is presented to architects, specifiers and others in the supply chain.

As Peter Caplehorn, chief executive of the Construction Products Association (CPA) put it:

'One of the painful lessons we learnt as a sector from the Grenfell disaster was that construction product standards must be made more robust and product information standardised and presented in a less ambiguous format. Regulatory frameworks need to be in place with an accompanying feedback loop and necessary sanctions to make sure products perform in a building as they are supposed to'.[14]

The CPA's Marketing Integrity Group has begun a programme to drive through change for higher ethical standards in product marketing information.[15] The programme has been informed by a survey (published in autumn 2019)[16] which confirmed that many architects and specifiers experienced difficulties in accessing the information needed to assess the performance of a product and to make informed decisions.

The group sets out two important principles in its survey findings:

- Construction product providers have a responsibility to make sure their product information is concise, complete, accurate and up-to-date.
- Marketers can help by ensuring all product information is provided in a clear way, written in plain English, accessible and easy to understand and compare.

In a video interview,[17] Peter Caplehorn talks about the role of PR, communications and marketing professionals in this issue:

> We wouldn't want to stop creativity. But to me the creativity should get people interested [in the product]. Then the description shows you the truth. It doesn't start to take you off down a path that misleads people into thinking that the product does something it doesn't do.

At the time of writing, a new voluntary code of conduct is being developed by the CPA to assist with the accurate communication of the testing, performance standards and benefits of building products. There is also talk of new competence standards that may extend to building product marketing and communications teams.

In the meantime, this is exactly the area where PR practitioners working to the professional standards set out by the CIPR[18] can add genuine value to product manufacturers and their customers, ensuring that our communications are conducted to the highest possible standards on issues of integrity, honesty, capability, capacity, competence, confidentiality, transparency and conflicts of interest.

Notes

1. The paid-earned-shared-owned (PESO) model describes the integration of PR and marketing communications channels.
2. Analysis of Vuelio media database, 19 September 2019.
3. Analysis of Vuelio media database, 19 September 2019.
4. UK Construction Week: https://www.ukconstructionweek.com/exhibit [Accessed 8 August 2020].
5. The Building Regulations, a set of Statutory Instruments which set minimum standards for design, construction and alterations to buildings in the UK, can be found at http://www.legislation.gov.uk/uksi/2010/2214/contents/made [Accessed 22 January 2020].
6. HM Government, Constructing 2025: https://assets.publishing.service.gov.uk/government/uploads/system/uploads/attachment_data/file/210099/bis-13-955-construction-2025-industrial-strategy.pdf [Accessed 8 August 2020].
7. London emissions targets: London Assembly policy on climate change mitigation: https://www.london.gov.uk/what-we-do/planning/london-plan/current-london-plan/london-plan-chapter-five-londons-response/policy-51-climate-change-mitigation [Accessed 8 August 2020].
8. BREEAM: https://www.breeam.com/ [Accessed 8 August 2020].
9. Part L of the Building Regulations: https://www.gov.uk/government/publications/conservation-of-fuel-and-power-approved-document-l [Accessed 22 January 2020].
10. Timber Expo is part of UK Construction Week, the country's largest construction event held at the NEC, Birmingham each autumn: https://www.ukconstructionweek.com/timber-expo [Accessed 8 August 2020].
11. RAL colours: https://en.wikipedia.org/wiki/RAL_colour_standard [Accessed 8 August 2020].
12. Hackitt Review, Independent Review of Building Regulations and Fire Safety 2018: https://www.gov.uk/government/collections/independent-review-of-building-regulations-and-fire-safety-hackitt-review [Accessed 8 August 2020].
13. Industry Safety Steering Group: https://www.gov.uk/government/publications/building-safety-industry-safety-steering-group-second-report-on-culture-change-in-the-construction-sector [Accessed 8 August 2020].
14. Designing Buildings Wiki, Construction Products Association call for evidence survey, 7 April 2019: https://www.designingbuildings.co.uk/wiki/Construction_Products_Association_call_for_evidence_survey [Accessed 8 August 2020].
15. CPA blog, Why we issued a Call for Evidence on construction product information, 1 April 2019: https://www.constructionproducts.org.uk/news-media-events/blog/2019/april/why-we-issued-a-call-for-evidence-on-construction-product-information/ [Accessed 8 August 2020].

16 CPA Construction Product Information Survey, 31 October 2019: https://www.con structionproducts.org.uk/publications/technical-and-regulatory/construction-product -information-survey/ [Accessed 8 August 2020].
17 Marketing in construction: an interview with Peter Caplehorn: https://www.lizmale. co.uk/lmc2/marketing-in-construction-an-interview-with-peter-caplehorn/ [Accessed 8 August 2020].
18 Chartered Institute of Public Relations Code of Conduct: https://cipr.co.uk/CIPR/A bout_Us/Governance_/CIPR_Code_of_Conduct.aspxg [Accessed 8 August 2020].

7 Communications for infrastructure

Successful stakeholder engagement on transport and infrastructure projects

Jo Field

Introduction

There has never been a more important time for civil engineers. Engineers shape the world around us, building the infrastructure and transport systems that enable countries and cities to grow and information and goods to travel around the world. Transport is the lifeblood of local economies, connecting people to jobs and vital services, adapting to population growth and enabling development of homes and communities. Infrastructure creates jobs, boosts economic growth and powers local economies – it is no surprise then, that major projects are forming such an important part of the UK's response to the impacts of the COVID-19 pandemic.

Government and local government stakeholders recognise the importance of investment in major infrastructure to drive recovery and growth across all regions of the UK, to boost connectivity, to get people back to work and to improve passenger experience.

There is also an acknowledgement of the need to mitigate the impacts of projects on local communities, both during construction and once major infrastructure projects are operational, and to create projects that benefit the communities they serve.

The purpose of PR for major transport and infrastructure programmes is two-fold.

First, there is a legal duty to consult the public. Indeed, there are particularly strict requirements about how to consult on major infrastructure projects in the planning system in England and Wales, via the Nationally Significant Infrastructure Projects (NSIPs) planning protocols.[1]

Under the Planning Act 2008 (as amended), a developer intending to construct an NSIP must obtain 'development consent'. The National Infrastructure Directorate of the Planning Inspectorate receives and examines applications for development consent.

In the case of transport NSIPs, after examining an application, the Planning Inspectorate makes a report and recommendation on the project to the Secretary of State for Transport. The Secretary of State for Transport then decides whether to grant or refuse development consent. If the decision is to give consent for a project to go ahead, the Secretary of State for Transport will make a Development Consent Order. This contains the consent and other authorisations (e.g., to purchase land compulsorily) which the developer needs to construct and operate the project.[2]

Given this process, there are obviously benefits both for the project and for an organisation's reputation in managing this well and preventing or managing subsequent legal challenges.

Second, given the nature of big civil engineering programmes, due to the lengthy construction times involved and the fact decision-making is usually taken at a Central Government level, getting stakeholders engaged in the process and building advocacy is essential to making major infrastructure projects happen.

Transport and infrastructure PR is all about engaging stakeholders and involving them on the journey.

This chapter looks at what we mean by stakeholder engagement and why it is important for transport and infrastructure projects. It also outlines the strategic approach to PR in this sector and some of the techniques used. It explains the importance of building advocacy and taking an innovative approach to engagement, in partnership with stakeholders and includes case studies to illustrate this. The chapter also describes how to measure success and summarises the author's tips for successful stakeholder engagement.

The main areas of practice in transport and infrastructure PR are community and stakeholder engagement, public affairs and media relations.

Why stakeholder engagement is important

Defining stakeholder engagement

Stakeholder engagement means building relationships with the communities and groups that have an interest in an infrastructure scheme. Tench and Yeomans[3] state that stakeholder interests may be 'direct or indirect …active or passive, known or unknown, recognised or unrecognised'.

In the context of transport and infrastructure projects, it is important for PR professionals to seek out all those stakeholders who are directly and indirectly impacted by the scheme, as well as those who may be unaware why they should have an interest.

A key objective of stakeholder engagement on transport and infrastructure projects is creating opportunities for decision makers to hear from the people that their decisions will affect – the stakeholders themselves.

It is important to engage a wide range of stakeholders in the planning and design of transport and infrastructure projects from the earliest possible stage. Ultimately, this means better policy decisions will be made because the people affected by the outcome of transport and infrastructure decisions are able to have their say on what will and won't work in their local community.

While programme directors and decision-makers may be tempted to limit the opportunities to consult and engage on a project moving at pace, in order to deliver the scheme, it is important to remember that transport and infrastructure is about people. Decision makers are planning, designing and delivering schemes for people, so personal engagement is crucial. PR professionals often need to be the voice of reason and to internally champion the need for customer engagement above and beyond what the law requires.

This is important because stakeholders are far more likely to support and cheerlead a project they have been involved in shaping from the earliest possible stage.

The need for continued case-making

Major transport and infrastructure projects tend to be long-term programmes. It can take years before decisions are taken to give projects the go-ahead. As an example, London's Crossrail was first proposed in the 1940s, with construction not starting until 2009; Heathrow Airport's expansion has been on the table since the 1960s and at the time of writing still remains unresolved, and High Speed 1 (the Channel Tunnel rail link) was first proposed in the 1970s, with construction not starting until 1998.

Once schemes get final approval, there can be years of construction disruption before the benefits are finally realised. Indeed, some communities will believe such schemes do not benefit them at all.

The role of PR, therefore, is crucial. This enables early engagement with trusted stakeholders to help clients and scheme promoters shape their ideas and develop better policy. Communications and engagement campaigns, where stakeholders carry content and amplify the scheme promoter's messages to influence the political environment, also build support for the scheme and protect and enhance its reputation.

Transport and infrastructure projects need stakeholder advocates, or cheerleaders, to continually make the case for a scheme, to protect the programme and enable it to sustain support across political cycles, including changes of local and national government. Political will plays a key part in decision-making on many transport and infrastructure schemes. In times of economic recession, big projects are easy to cancel or delay. This is exactly what happened to Crossrail – the scheme was rejected by Parliament during the recession in the early 1990s.[4] Crossrail eventually came to fruition because UK business backed it and was willing to campaign for it and pay its share of the costs.

While cancelling big schemes is arguably bad for the economy, as new infrastructure creates jobs and boosts economic growth, political priorities and views on the costs and benefits of a project may change with a new administration, a change in leadership or increased public funding pressures.

This means that complacency is not an option. There is always the need to make the long-term, continued political case for transport and infrastructure projects in the national interest. Case-making is needed from the start to finish of the programme and often even once the project is operational.

110 Jo Field

Figure 7.1 sets out the various stages of major transport and infrastructure projects from initial concept through to detailed design, enabling works, construction and operations. It shows the role of stakeholders and why it is important to engage and mobilise them at each different stage of the project.

Communicating the social value of infrastructure

Increasingly, PR professionals in the transport and infrastructure sectors need to show the social value of infrastructure as part of their continued case-making campaign.

While job creation is one of the key social value 'proof points' of infrastructure projects, there is also a focus on the 'legacy' that schemes will leave behind in the communities they impact.

This includes the legacy in terms of community investment – funding 'pots' made available to local communities – and projects delivered in partnership with local people by public sector clients, developers and their works contractors. These are projects and programmes that develop capacity, health and wellbeing in local communities and upskill local people.

The Public Services (Social Value Act) 2012 sets out a legal duty for public sector organisations to consider how procurement could improve the social, economic and environmental wellbeing of the relevant area.[5]

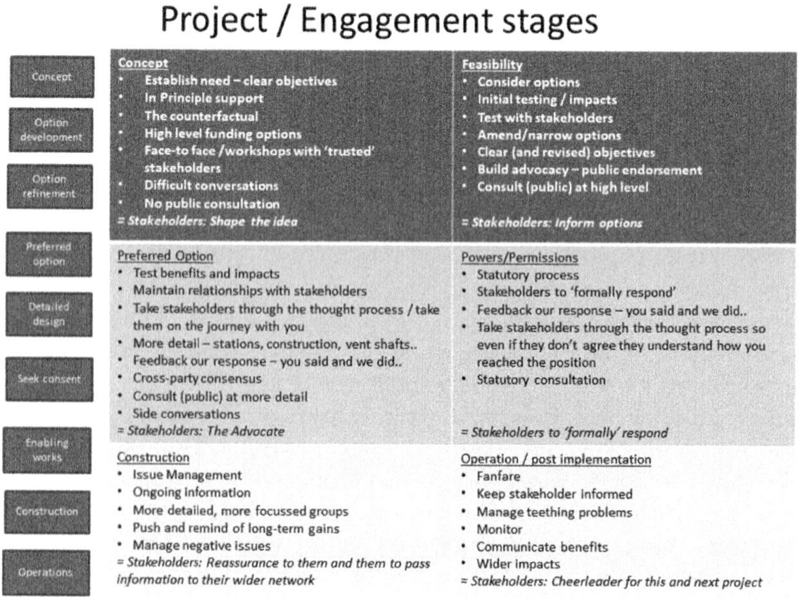

Project / Engagement stages

Stage		
Concept	**Concept** • Establish need – clear objectives • In Principle support • The counterfactual • High level funding options • Face-to face /workshops with 'trusted' stakeholders • Difficult conversations • No public consultation = *Stakeholders: Shape the idea*	**Feasibility** • Consider options • Initial testing / impacts • Test with stakeholders • Amend/narrow options • Clear (and revised) objectives • Build advocacy – public endorsement • Consult (public) at high level = *Stakeholders: Inform options*
Option development		
Option refinement		
Preferred option	**Preferred Option** • Test benefits and impacts • Maintain relationships with stakeholders • Take stakeholders through the thought process / take them on the journey with you • More detail – stations, construction, vent shafts.. • Feedback our response – you said and we did.. • Cross-party consensus • Consult (public) at more detail • Side conversations = *Stakeholders: The Advocate*	**Powers/Permissions** • Statutory process • Stakeholders to 'formally respond' • Feedback our response – you said and we did.. • Take stakeholders through the thought process so even if they don't agree they understand how you reached the position • Statutory consultation = *Stakeholders to 'formally' respond*
Detailed design		
Seek consent		
Enabling works	**Construction** • Issue Management • Ongoing information • More detailed, more focussed groups • Push and remind of long-term gains • Manage negative issues = *Stakeholders: Reassurance to them and them to pass information to their wider network*	**Operation / post implementation** • Fanfare • Keep stakeholder informed • Manage teething problems • Monitor • Communicate benefits • Wider impacts = *Stakeholders: Cheerleader for this and next project*
Construction		
Operations		

Source: © Esme Yuill, Transport for London

Figure 7.1 Key stages for engagement in major transport and infrastructure projects.

PR professionals need to place themselves right at the heart of shaping and telling the social value stories of infrastructure projects. This can include working with stakeholders to explain why transport matters to people and using case studies to bring a project to life and support the story.

Stakeholder and community engagement professionals forge relationships with the local community, identifying the legacy outcomes the community wants to achieve and working in partnership to deliver them. In theory, the better and stronger the relationships develop, the greater the social value outcomes will be, as the community engagement professional has gained a deep understanding of community needs.

When legacy is a top priority on transport and infrastructure schemes, the social value story should be a leading part of the project's narrative. This is because it is hard to argue against community investment and job creation. Even if the scheme is not popular, the legacy can be.

Relevant policy and legislation

There is not always a legal requirement to consult on private sector transport and infrastructure schemes, as explained later, but it is considered best practice.

There will be a statutory consultation process once a planning application is submitted (and, in the case of NSPIs, prior to submission). But for private sector infrastructure providers there may also be a case for additional earlier consultation activities in order to understand the audiences and to ensure good policy making.

For the public sector, there are several policy recommendations and pieces of legislation that prescribe varying levels of duty for government, local authorities and developers to consult on transport and infrastructure schemes. These include:

Duty to consult

The duty to consult is the principle that any public authority should exercise fairness in its practice, as its decisions need to be able to stand up to public scrutiny. Such scrutiny could include judicial review, where a judge reviews the lawfulness of a public body's decision, or an appeal to a higher court.

The Government has a Code of Practice 2008[6] which lays out seven criteria to consider during a consultation. These include guidance on when to consult – when there is scope to influence the policy outcome, consultation length (usually 12 weeks) and guidance on scope, accessibility, analysis and feedback. But not all public authorities follow the Government's code of practice. Many local authorities have their own policy, and the onus is very much on the public authority to decide where, when and how to consult.

Duty to promote equality

The Public Sector Equality Duty[7] was created by the Equality Act 2010 and requires public bodies to:

- Eliminate unlawful discrimination, harassment and victimisation and other conduct prohibited by the Act.
- Advance equality of opportunity between people who share a protected characteristic and those who do not.
- Foster good relations between people who share a protected characteristic and those who do not.

Equality Impact Assessment (EqIA) is one of the tools used to determine whether a policy has the ability to affect protected groups differently. It is best practice to carry out an EqIA as part of the consultation process in the development of any transport and infrastructure scheme, no matter how small, to determine the scheme's impact on protected groups. This helps the scheme's developers to get it right for everyone the first time and avoids legal challenge and expensive retrofitting later. PR has an important role to play in ensuring policy makers can engage with protected groups, gather evidence to inform the policy and make sure stakeholders' views are responded to.

Duty to consult the local community

While there is not always a legal duty to consult on transport and infrastructure schemes, there is on major projects. As mentioned earlier, the Local Planning Act 2008 introduced legal requirements for NSIPs, such as major roads and airport expansions, to formally consult all statutory bodies, local authorities, communities and affected persons at the pre-application stage.

The Localism Act 2011 transferred the decision-making powers for NSIPs to the Secretary of State.[8]

Development Consent Order (DCO) is the mechanism for seeking planning approval for NSIPs. The application is made to the Planning Inspectorate, who considers the application and makes a recommendation to the Secretary of State. The Secretary of State then gives final approval.[9]

With a DCO, the emphasis is on engagement and consultation with residents, stakeholders and other affected groups before the application is submitted. There are continued opportunities for groups to influence the planning process over the duration of the project.

Large, national infrastructure projects that significantly impact many people and stakeholders require planning consent from Parliament by way of a Hybrid Bill.

Only 13 Hybrid Bills have been introduced in Parliament since 1984[10] including the Channel Tunnel and Crossrail. At the time of writing, the High Speed Rail (West Midlands – Crewe) Hybrid Bill is currently passing through Parliament.

During the Hybrid Bill process, both the House of Commons and House of Lords debate the outcome. Opponents of the Bill, including members of the public, are able to submit petitions (their case against the Bill). Some individuals and groups will be invited to present to a Select Committee.

The engagement with all affected parties is ongoing throughout the Hybrid Bill process.

The role of PR during the planning process

PR plays a central role in both the DCO and Hybrid Bill process; it is vital in getting to the planning approvals stage in the first place. Figure 7.1 shows how important it is to build stakeholder support and for stakeholders' views to shape the initial concept.

PR professionals work with communities and stakeholders to make sure that they are properly briefed on a proposed scheme, to address concerns – often before they are voiced publicly – and to build support as the project progresses to seeking permissions and powers.

Public affairs professionals work with stakeholders and elected representatives to build support and ensure political approval of the scheme. They also work with stakeholders to influence the surrounding political environment and amplify messages in the media. Media relations plays a central role in shaping and communicating the narrative more widely.

Strategic approach

Transport and infrastructure projects are long-term and must progress through a number of stages before receiving final approval. The construction process is also likely to be lengthy. Therefore, long-term continued case-making is vital in taking a strategic approach to PR for transport and infrastructure schemes and should be central to the project's overall objectives. Any scheme that fails to consider the value of PR or includes it as an after-thought is unlikely to be viewed favourably by decision makers.

Like the projects themselves, the case-making campaign is complex. It needs to be well-planned and orchestrated in partnership with stakeholders from an early stage. It needs to be regularly reviewed and refined, with clear, measurable objectives. Major projects need to be supported and driven by large-scale communications and engagement campaigns for the duration of the project and beyond.

Stakeholder mapping

The first stage of stakeholder engagement is to carry out a thorough mapping exercise to identify and understand the stakeholders and to research their views of the project.

It is important to research and understand all the project's stakeholders, including those the scheme directly and indirectly impacts, as well as those groups who may be unaware of why they should have an interest.

It is best practice to plot stakeholders on a Power and Interest grid (see Figure 7.2) and understand their positions of power, interest and influence over

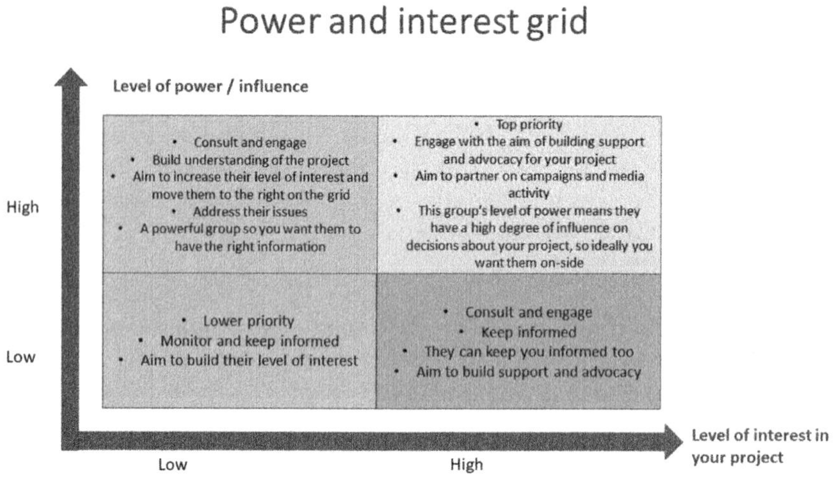

114 Jo Field

Power and interest grid

	Low interest	High interest
High power	• Consult and engage • Build understanding of the project • Aim to increase their level of interest and move them to the right on the grid • Address their issues • A powerful group so you want them to have the right information	• Top priority • Engage with the aim of building support and advocacy for your project • Aim to partner on campaigns and media activity • This group's level of power means they have a high degree of influence on decisions about your project, so ideally you want them on-side
Low power	• Lower priority • Monitor and keep informed • Aim to build their level of interest	• Consult and engage • Keep informed • They can keep you informed too • Aim to build support and advocacy

Axes: Level of power / influence (vertical); Level of interest in your project (horizontal).

Figure 7.2 Power and interest grid.

the project and their ability to sway decision makers' or public opinion. For example, if the project is an NSIP, then the Secretary of State is high in both power and interest as the ultimate decision maker. A local campaign group in the impacted area will have high interest but significantly less power.

In mapping stakeholders, it is important to think about the main audiences to engage, who trusts them and who they trust and will listen to. Consider existing stakeholder relationships and how these can be helpful in building support or reaching out to new groups.

As part of the mapping exercise, research which stakeholders have been active on this policy or project before – with the organisation, in the media or online – and find out their current views of the project and their key issues. PR professionals usually work in partnership with policy or technical colleagues to develop a detailed understanding of the stakeholder landscape.

It is also important to think about how the different stakeholder groups influence each other. While a local campaign group has less power than the Secretary of State, a well-mobilised and well-argued campaign against a scheme has significant ability to sway public and political opinion. It is also important to think about which stakeholder groups may be feeding questions to elected representatives, which groups are talking to the media and which stakeholders are connected in surprising ways. An individual campaigner may at first glance appear to have limited influence, but he or she may be well plugged in to the media, have a large following on social media or have friends in positions of authority. A savvy PR professional will be aware of all these interactions and will work with stakeholders and the media to build understanding and support of the infrastructure project.

Understanding stakeholders' positions on the grid and how they relate to each other will help to determine the level of information and engagement that stakeholders need. It is important to remember stakeholders' positions on the grid change over time, so it is important to re-examine the grid at regular intervals. Ideally, bring as many stakeholders as possible to a position of 'high interest', ready to be engaged with the aim of securing advocacy and open to joint campaigning and media activity.

Once priority stakeholders have been identified, their current sentiment should be considered. It is then possible to move to a supportive sentiment throughout the communications and engagement activity.

Why it is important to capture the views of a wide range of stakeholders

Aside from statutory duties to consult and to ensure new infrastructure does not disadvantage protected groups, there are many important reasons to ensure a range of stakeholder voices inform the development of new transport and infrastructure projects.

It is important to have the conversation and not to avoid engagement with groups who have a reputation for being staunchly opposed to proposals or combative on issues associated with the project. This may sound obvious, but PR professionals often have to persuade technical colleagues and programme directors of the need to engage far and wide to ensure that all views are captured and to enable relationship building that can result in a positive change in stakeholders' attitudes. Attempting to minimise or avoid engagement with any group will lead to problems later, as stakeholders will become frustrated at a lack of dialogue.

For example, engaging with disabled people's organisations and involving them early in shaping the scheme will ensure that the developer gets things right first time, rather than having to spend money on expensive retrofitting later. PR professionals should encourage scheme promoters to carry out EqIAs for every new project. Getting it right for disabled people means getting it right for everyone. After all, many adjustments that transport schemes make for disabled people, such as wheelchair access ramps, actually make life easier for everyone.

Given the long-term timescales of transport infrastructure from initial concept to construction and operations, young people are a vital audience. Young people are the future users of major projects, so it is right they have a say in decisions that will affect them.

Equally, young people are the future engineering and infrastructure workforce, and we need to inspire their interest and encourage them to join the industry. Government figures show an infrastructure skills gap, with rail needing an extra 50,000 people by 2033 and roads needing another 41,000 people by 2025.[11] This means there are not enough workers with the skills needed to deliver the government's pipeline of transport infrastructure projects. This coupled with the massive under-representation of women in the transport workforce (only one-fifth of transport workers are women),[12] means the industry needs to improve its engagement with young people to create wider awareness of the industry and encourage more people to consider transport careers.

An example of best practice in engaging young people is provided in the case study below.

Box 7.1 Transport for London (TfL) Youth Panel

TfL's Youth Panel was set up in 2009 to give young people a direct voice in policy-making.

The panel is made up of around 25 young people aged between 16 and 25 who regularly travel in London. The panel recruits every January and holds elections for executive committee roles, including chair and vice-chair, each April. The panel provides development opportunities and inspires young people from across London's diverse communities to contribute to shaping a network that meets their needs.

The panel has helped shape many key policy initiatives and campaigns, including:

- The Mayor's Transport Strategy
- TfL's equality commitments
- Road safety campaigns
- Year of Engineering

Members also formally submit consultation responses, representing the voice of young Londoners. The Youth Select Committee, which mirrors the UK Parliament's Select Committee process, called on all passenger transport providers to set up youth panels based on the TfL model. Transport providers around the world have set up similar youth involvement processes.

Public relations models and extent of two-way engagement

Engagement fits with Grunig and Hunt's (1984) two-way symmetric model of PR. This is 'a model where PR listens to the public and changes according to the publics' needs and desires'.[13]

Engagement is the optimum method of stakeholder communication that PR professionals should aim for. This is because it is a two-way, continuous dialogue and involves proposals, plans and communications and engagement methods being shaped and refined in equal partnership with stakeholders. Engagement is superior to 'involvement', which is regarded as 'asymmetric', rather than 'symmetric'.

While engagement is the optimum method of communication, it is more about quality than quantity, so it will never be sufficient on its own. As explained previously, most transport and infrastructure schemes require an element of large-scale public consultation as part of the statutory decision-making process. As projects progress through construction, there will always be an element of public information and instruction required alongside more detailed engagement.

Developing policy and plans with stakeholders

As Figure 7.1 shows, the early stages of the project are about establishing need and building support from trusted stakeholders. This stage is about stakeholders shaping an idea, rather than wide-scale public consultation.

Some projects are shaped by stakeholders presenting an idea to public sector clients and then lobbying for the scheme to progress. One such scheme is the Rotherhithe Bridge, a proposed walking and cycling bridge across the River Thames. The bridge was initially proposed by sustainable transport charity, Sustrans. It was adopted by the Mayor of London, Sadiq Khan, and included in his 2018 transport strategy. The bridge progressed through the early design and options stages but was cancelled in 2019 due to a lack of funds. This shows that even when there is political will and stakeholder support, initiatives don't always progress.

Generally public sector clients develop the policy solutions needed to solve a particular challenge such as capacity, population growth, air quality or congestion, and then seek the stakeholder support to deliver it.

It is best practice, however, to develop policy solutions and plans together with stakeholders, working together to achieve shared objectives. This is the beginning of the stakeholder engagement process.

Many stakeholder or representative groups (including membership bodies, trade associations and campaigning charities) will have in-house transport policy expertise that clients and scheme promoters should want to access during the early stages of policy development and to continue the conversation throughout the project stages. Importantly, these groups represent the customer or end-user, who can be trusted advocates to help get the proposed scheme right for their client group.

PR plays a central role in the policy development phase. PR professionals develop and nurture stakeholder relationships over time and gain a deeper understanding of their stakeholders' policy positions, as well as planning tactics and enabling the trusted conversations to take place. As the project engagement stages develop, PR professionals work with stakeholder groups to shape and develop the campaign, amplify its messages and build support among the general public.

There are many different types of stakeholder. Box 7.2 below shows examples of different types of stakeholder groups.

> **Box 7.2 Different types of stakeholder groups**
>
> Parliamentarians – MPs and peers
> Councillors
> Assembly members
> Mayors
> Civil servants
> Ministers
> Government departments
> Local councils
> Community groups
> Residents' associations
> Employees
> Customers
> Suppliers
> Media
> Opinion leaders and influencers
> Membership bodies/trade associations
> Business and business representative groups
> Industry bodies
> Charities and campaign groups
> Environmental groups
> Equalities groups
> Disabled people's organisations
> Statutory watchdogs/consumer groups
> Young people/youth groups
> And many more…

Advocacy and its role in protecting reputation

When stakeholders amplify or cheerlead messages, it is known as advocacy.

Stakeholder advocacy is crucial to enable transport and infrastructure projects to weather political cycles and to protect their reputations as they progress through the design, planning and construction stages.

Stakeholder advocates can help make the case for and shape transport and infrastructure schemes. A chorus of voices from a range of stakeholder groups adds credibility and impact to the scheme and will persuade decision makers of its value.

But advocacy does not happen overnight. Advocacy is the result of time, energy and effort spent in building and nurturing long-term relationships with stakeholders and working together to achieve shared objectives. Stakeholders are more likely to support policies that they have helped to shape and when they feel they have been listened to. That's why it is important to engage early and often and to ensure that stakeholders have real opportunities to influence proposals. Engagement is an ongoing, iterative process and should never be a 'tick-box' exercise. Only high-quality, continuous, meaningful engagement will produce advocacy.

PR professionals should aim to build a coalition of stakeholder advocates that are willing to campaign in support of the project. This continual case-making is what builds political and public support for planning approval and protects the reputation of the project as it progresses through the construction phases through to operations.

An example of stakeholder advocacy on a scheme that has faced substantial opposition is the Stonehenge Tunnel – a proposed £1.7bn road tunnel and upgrade of the A3030 next to a World Heritage Site and one of the world's most famous circles of standing stones.

Historic England, the National Trust and English Heritage announced their support for the tunnel after Highways England listened to their views and changed its preferred route. The Secretary of Transport has now confirmed that the project will go ahead.[14]

A Chartered Institute of Public Relations interview[15] with the chief executive of Heathrow, John Holland-Kaye, highlights the role of Heathrow's communications team in securing the stakeholder support and advocacy that was vital in the airport receiving the go-ahead for its third runway.

In the interview, Holland-Kaye describes PR as 'vital to managing our reputation' and revealed his thoughts on the distinguishing characteristics that PR professionals bring to the table, including their skills in targeting appropriate stakeholders and ensuring the accurate communication of facts.

It is not advisable to scale back communications once planning permission is received. Construction is a lengthy, disruptive process. Big projects naturally attract public and media attention, and there will be no shortage of negative stories. Aside from the statutory communications and public information needed to accompany the construction, it is essential to have supportive stakeholders continually champion the scheme and amplify its benefits. This is the single biggest factor in protecting the project's reputation as it progresses. Indeed, continued advocacy is essential to making sure the project does progress.

The case-making shouldn't stop once the project is operational. As Figure 7.1 shows, stakeholder advocates can continue to communicate the benefits and cheerlead the next scheme.

Advocacy as a two-way process

Rather than thinking about 'managing' stakeholders and 'defending' a project's reputation, PR professionals should think about how they can deliver top-quality meaningful engagement in partnership with stakeholders to achieve shared objectives. This work, sustained over time, is more likely to lead to advocacy.

120 Jo Field

Like stakeholder engagement, advocacy is a two-way process. PR professionals need to advocate for stakeholders, championing their needs internally. Communications teams should make sure that stakeholders' views are heard by the right people at the right time and facilitate dialogue with decision makers. For example, it is important to make sure that commitments are followed up on and to explain which options have been looked at and what the alternatives are if a solution proposed by a stakeholder isn't possible. Stakeholders need to know that someone is looking after their interests internally.

Even when stakeholders do not like a scheme and will never support it, they will appreciate the continuous dialogue and relationship with the PR team. This could lead to positive feedback about the quality of engagement, which is equally as valuable as advocacy for an organisation's reputation.

Techniques and channels

Project engagement needs to be participatory, interactive and tailored to the audience and the project stage as shown in Figure 7.1. For example, while a roundtable discussion may be the most effective method of gaining early views from trusted stakeholders, it will not be the most effective way of reaching large volumes of people at the statutory consultation stage.

It's important to ask stakeholders how they would like to be engaged with and, as far as possible, meet them on their terms and to their timescales.

A large-scale, continued case-making campaign will use a range of different techniques and channels. For example, parliamentarians may be engaged through one-to-one meetings and briefing papers tailored to their constituency interests. Residents and community groups are more likely to be engaged through online information, social media, broadcast media and events such as forums, public meetings and town hall events.

Media and social media can be used to reach a wider pool of people. In this instance, media and stakeholders (representative groups) are both an audience and a channel. They are a primary audience for PR activities, but they are also a channel to help amplify the campaign's reach to a wider audience. This is very important when carrying out a campaign for a high-profile infrastructure project.

Developing communications and engagement plans with stakeholders

Communications and engagement plans should be developed with input from stakeholders, alongside policy proposals. This is true throughout the project engagement stages, from initial concept and design all the way through to construction and operations.

Failure to develop communications and engagement in partnership with stakeholders or pre-empt the outcome means that the engagement will have limited chance of success.

Communications and engagement on infrastructure projects should always be a two-way, continuous conversation of shaping, developing and refining messages. This mirrors the project stages – with continual feedback, shaping and refining of proposals.

An example of best practice in developing communications in partnership with stakeholders is provided in the case study below.

Box 7.3 HS2 and Costain Skanska joint venture – community hoardings project

HS2 and its enabling works contractor, Costain Skanska joint venture (CSJV), carried out HS2's first high profile community engagement hoarding project in Drummond Street, Euston, in London. The project engaged and informed the local traders from the outset.

Drummond Street is a diverse street near Euston Station in London, occupied by many businesses. Demolition works for the new railway in the surrounding area meant that the Drummond Street traders were directly affected by the HS2 works.

The HS2 Act provided a series of assurances to the traders, including one related to hoardings. This gave assurance that if hoardings affected access to or visibility of businesses on Drummond Street, notices would be displayed confirming that they remained open and with directions showing how to access them.

There was a desire from HS2, CSJV and the community to produce an artwork hoarding that would look slick and show that businesses were open as usual. But, given the timetable of early works, and the extent of community engagement needed, an interim solution was provided. This meant a period of plain white hoarding, followed by an interim signage and wayfinding solution, before the commissioned artwork was delivered.

An extensive community engagement process took place to reach the final hoarding designs:

- First round of engagement: HS2 and CSJV asked traders what they wanted to see and what would best represent Drummond Street. (Feedback enabled the design team to create initial concepts).
- Commissioning of photographers and photoshoot.
- Second round of engagement: setting timeframes for photoshoots and gaining verbal consents.
- Third round of engagement: traders were engaged on the results of the photoshoot for final, signed consent. (During this process, there was constant feedback to the design team to capture comments).
- Fourth round of engagement: the designer produced five draft concepts. These were presented to all the traders for their feedback.
- The choice was reduced to two options:

- Focus groups – with a self-selected group of traders who felt particular ownership of the street. (Everyone's views had to be carefully balanced to reach a final decision, while also taking into account HS2's own organisational values).
- Engagement with the local authority – the London Borough of Camden – to obtain approval and sign-off of the artwork.

All engagements were captured in the communications plan. This is good practice and shows transparency.

The final hoarding produced was a mix of artwork that captured the identity of the traders themselves and highlighted the diversity of the area. HS2 and CSJV received positive feedback from the community.

Some lessons learned by HS2 and CSJV are useful for other projects undertaking community hoardings work:

- Minimise the time between plain hoarding and commissioned work.
- The procedure with the local authority must be identified at the start of the project.
- It is important to balance everyone's views carefully, while also considering the organisation's own values.
- Flexibility is crucial when engaging with busy traders.
- The importance of regular communications – real community engagement takes time and is not a tick-box exercise.

Figure 7.3 Examples of the HS2 hoardings on Drummond Street.

Communications for infrastructure 123

Figure 7.4 HS2 hoardings on Drummond Street.

Making sure that all voices are heard

It is important for PR professionals to ensure that proposed transport and infrastructure schemes reach so-called 'hard-to-reach' groups as part of the consultation and engagement process.

But it is worth thinking about what we mean by 'hard to reach'. Generally, groups are only hard to reach if PR professionals have not tried hard enough to reach them. Some communities may be difficult to reach by traditional communication and engagement methods, whether it is because of language, cultural background, geographic location or lack of access to technology. That does not mean that these communities are difficult to reach as such. Every community is easy to reach for someone. That is why it is important for PR professionals working on transport and infrastructure schemes to find and partner with local groups that can help access everyone who needs to be heard, for example, faith groups or organisations working with older people.

Box 7.4 Crossrail 2 mobile community engagement unit

One example of best practice in reaching groups who are traditionally less likely to attend consultation events is Crossrail 2's mobile community engagement unit.

In 2015, TfL carried out a public consultation on Crossrail 2, a proposed railway linking the national rail networks in Surrey and Hertfordshire via an underground tunnel through London. The organisation held 72 drop-in events along the proposed route to attract interest and engage the public in the areas where Crossrail 2 would have an impact.

The Crossrail 2 proposals spanned:

- A significant geographic area, with over 70 kilometres of new tunnels and 12–14 new stations in the densely populated central section of the route and serving 34 stations in Surrey and Hertfordshire.
- Over 20 MP constituencies, 19 local authorities (including 14 unitary London Boroughs, two county councils and one regional park authority).
- Over 200,000 properties within 250m of the proposed tunnels or worksites.

In many areas, events were hosted in traditional spaces such as libraries, community halls and churches. But TfL wanted to balance these more traditional venues with a mobile unit that could be located in the heart of the communities. This enabled the organisation to better target areas of high footfall and those areas near to where the proposals would have most impact.

The mobile unit was a converted long wheel based transit van. The van was wrapped in Crossrail 2 branding including the logo, route map and website address. Consultation staff wore Crossrail 2 branded tabards. These features meant the van was visible and enabled the consultation team to attract passers-by to the events.

Internally, the van was kitted out with generic and location specific information panels (which could be changed according to location), leaflet holders and a large screen TV. Fixed iPads provided staff with quick and easy access to the Crossrail 2 interactive map and the website, so they could show factsheets and maps to explain proposals and answer questions.

The strong Crossrail 2 branding on the van complemented the integrated advertising campaign that promoted the scheme. The van acted as a mobile billboard as it travelled from location to location across Herts, Surrey and central London.

A quick Survey Monkey questionnaire enabled staff hosting the events to provide feedback at the end of their shifts. Data was collected on numbers of people consulted, their interest in the scheme, the type of issues raised and any questions they couldn't answer. This helped TfL understand the emerging issues and identify the need for additional information or lines to take.

Taking the consultation out onto the streets meant that Crossrail 2 could capture the views of those who are traditionally less likely to attend consultation events. This method attracted around two-thirds of the almost 13,000 people that attended engagement events.

Using a mobile unit meant that TfL could be flexible on location and did not have to rely on using village halls or other public spaces that may not be in the ideal location. The van enabled TfL to hold engagement events near the sites they were consulting upon, for example, small residential streets and parks.

Locating the van in areas of high footfall helped to raise awareness of Crossrail 2. It attracted thousands of passers-by that would not have been reached through more traditional fixed venues.

But the van needed to be easy to park on the street or another hardstanding space, so it could not be too big. The van chosen could accommodate up to 10–12 people at a time – but any more and space became an issue. Where space and weather allowed, TfL were able to expand the engagement events outside of the bus. A mobile unit of this size should supplement, rather than replace, use of more traditional larger venues.

Figure 7.5 Crossrail 2 mobile community engagement unit.

126 Jo Field

Figure 7.6 Crossrail 2 mobile community engagement unit.

Innovative engagement

For stakeholder engagement to be successful, elements of innovation are incredibly helpful. This could include partnering with stakeholders to deliver creative engagement solutions such as industry 'firsts'. PR can also set the direction on innovation that is needed from the market to address policy challenges. The case study below outlines Transport for London's award-winning buggy summit and the associated campaign. The campaign was PR-led but sparked industry innovation among buggy retailers and manufacturers.

Box 7.5 Transport for London buggy summit

Transport for London's (TfL) award-winning buggy summit brought together a range of stakeholders with competing views to discuss and overcome policy challenges. The first of its kind in the UK, the buggy summit was designed to make travelling around London easier for everyone. The event brought together buggy manufacturers, retailers, parents' groups, disabled people's organisations and industry experts. Each group had a seat at the table to discuss the conflict that sometimes happens over the wheelchair priority space on London's buses. All sides were able to discuss their personal experiences with TfL senior leaders and policy-makers.

The summit was not about changing the policy, because the law gives wheelchair users priority for wheelchair spaces on buses. But the event shows best practice in engaging stakeholders to shape communications and develop joint solutions to policy challenges.

The buggy summit formed part of a wider campaign to help buggy and wheelchair users better navigate London's transport system. The campaign was further developed based on stakeholders' feedback at the summit. Key outputs included:

- Manufacturers and parents' groups asked to consider how everyone could work together to help overcome the challenges and shape future buggy designs.
- Retailers such as Amazon, Babies 'R' Us, Tesco, Mothercare, Argos and John Lewis committed to reviewing their sales strategies to promote smaller buggies to those using planning on using public transport.
- Agreement for a 'best buggy for public transport vote' – highlighting to customers the benefits of buying small, light-weight, foldable buggies to customers and profiling the best ones.
- Development of a 'kitemark' recognition scheme for public transport friendly buggies for vote winners.
- A bespoke parenting hub created on the TfL website with advice and travel tips.
- Production of an advice leaflet for people using buggies on TfL services. This was distributed through Mothercare's London-based stores, along with advisory information to store managers, encouraging them to raise awareness of the benefits of smaller buggies.
- Creation of a 'top tips' video for those travelling with babies and young children, which was viewed over 100,000 times.
- Enhancement of wheelchair priority spaces.
- The size of the priority space on buses was expanded where possible.
- The campaign was presented to the managing directors of the UK's bus operators, influencing the approach to the conflict around the wheelchair priority space nationally.
- The campaign secured advocacy from the disability sector in recognition of TfL's commitment to tackling this issue.

The campaign was named Public Sector campaign of the year in the 2016 Public Affairs awards. It built longstanding relationships with the parenting sector, which enabled TfL to secure a partnership with Mothercare and Baby Zen – best buggy vote winners – worth over £100K for its Baby on Board badge. This partnership channel allowed TfL to reach 130,000 pregnant women each year with information about the buggy priority space on buses.

Measuring success

SMART objectives

To measure the success of stakeholder engagement activities for transport and infrastructure schemes, it is advisable to create SMART (specific, measurable, achievable, relevant and time-bound) objectives to build effective relationships with key stakeholders and to track progress over time. Objectives could include a commitment to:

- Improve understanding of a project
- Secure stakeholder support/advocacy
- Minimise risks to the project's reputation
- Secure media coverage to get the key messages across
- Deliver better policy as a result of effective stakeholder engagement

Necessary data

Carry out regular surveys to gain key stakeholders' views of your communications and engagement activities and their sentiments about the project. A benchmarking survey should be carried out at the start of each project and repeated every year. Engagement activities are considered successful if there is a measurable increase in the quality of engagement and sentiment about the project over time.

Once engagement activities are underway, quantitative data can be collected and reported. This includes numbers of stakeholders attending meetings and events, responding to surveys, opening emails or engaging on social media. But it is important to examine the nuances behind this data. For example, 60 people attending an engagement event may sound good. But did the activity reach the intended audience? Did the desired range of stakeholder groups and levels of seniority go to the event?

Feedback forms and mini-surveys should be carried out following meetings, roundtables and forums to capture stakeholders' views and provide insights to inform the development of future communications and engagement activities and messaging.

Media coverage and stakeholder amplification of messages, where stakeholders carry key messages in their newsletters or on their websites, should be tracked and collated in a coverage book for reporting back to the project team.

Outputs, outtakes and outcomes

When measuring PR activities for transport and infrastructure schemes, it is important to recognise the difference between outputs, outtakes and outcomes.

An example of an output is a roundtable that engages ten key stakeholders to help shape the idea for a new transport scheme. The output is the 'output' of the communications activity, such as an event or a press release, and the number of stakeholders engaged or the amount of coverage in the press.

The outtake in this case is how the stakeholders reacted to the engagement activity: their view of the project, their reaction to the scheme being mentioned in the media and whether they share content positively or negatively.

The outcome is what happened as a result of the engagement activity – for example, the attitudinal shift of stakeholders to more supportive or more opposed.

Conclusion

The key to successful stakeholder engagement on transport and infrastructure is working in partnership with stakeholders to deliver shared objectives. This means engaging stakeholders early and often and involving them in the journey over the long term.

Stakeholder advocacy is absolutely essential in protecting a project's reputation and enabling it to sustain support across political cycles. Advocacy is the result of working in partnership with stakeholders and delivering continuous, high-quality, meaningful engagement over a sustained period. It is important to remember that advocacy is a two-way process. Successful stakeholder engagement on transport schemes involves championing your stakeholders' views.

Box 7.6 Author's tips for successful stakeholder engagement

Set clear objectives for your engagement.

Identify your audiences and the stakeholders that represent them.

Map your stakeholders and prioritise engagement.

Have the conversation.

Engage early and often.

Build advocacy for the project.

Be a champion for your stakeholder.

Evaluate your impact on stakeholder attitudes over time.

Notes

1 Planning Inspectorate, National Infrastructure Planning: https://infrastructure.planninginspectorate.gov.uk/ [Accessed 8 August 2020].
2 Department for Transport, Guidance on nationally significant infrastructure projects in the transport sector, updated 16 April 2020: https://www.gov.uk/government/publications/nationally-significant-transport-infrastructure-projects/nationally-significant-infrastructure-projects-in-the-transport-sector [Accessed 8 August 2020].
3 Tench, R. & Yeomans, L. (2006) *Exploring Public Relations: Global Strategic Communication*. 4th edn. Harlow: Pearson Education Ltd.
4 Crossrail website: http://www.crossrail.co.uk/route/crossrail-from-its-early-beginnings [Accessed 28 January 2020].

5. https://assets.publishing.service.gov.uk/government/uploads/system/uploads/attachment_data/file/690780/Commissioner_Guidance_V3.8.pdf [Accessed 28 January 2020].
6. https://assets.publishing.service.gov.uk/government/uploads/system/uploads/attachment_data/file/100807/file47158.pdf [Accessed 28 January 2020].
7. https://www.equalityhumanrights.com/en/advice-and-guidance/public-sector-equality-duty [Accessed 28 January 2020].
8. https://infrastructure.planninginspectorate.gov.uk/legislation-and-advice/legislation/ [Accessed 28 January 2020].
9. https://infrastructure.planninginspectorate.gov.uk/application-process/the-process/ [Accessed 28 January 2020].
10. https://researchbriefings.parliament.uk/ResearchBriefing/Summary/SN06736 [Accessed 28 January 2020].
11. Strategic Transport Apprenticeship Taskforce (July, 2018) *Transport Infrastructure Skills Strategy, two years on* (https://assets.publishing.service.gov.uk/government/uploads/system/uploads/attachment_data/file/727052/transport-infrastructure-skills-strategy-two-years-on.pdf) [Accessed 28 January 2020].
12. Trans.info, Women represent about one fifth of transport workers, which EU country has the highest share in the sector? 8 March 2018: https://trans.info/en/women-already-one-fifth-transport-industry-eu-country-highest-share-sector-84140 [Accessed 8 August 2020].
13. Description of Grunig and Hunt's model in pp.152 Tench, R. and Yeomans, L. (2006) *Exploring Public Relations: Global Strategic Communication*. 4th edn. Harlow: Pearson Education Ltd.
14. Highways England, A303 Amesbury to Berwick Down (Stonehenge): https://highwaysengland.co.uk/a303-stonehenge-home/ [Accessed 8 August 2020].
15. https://newsroom.cipr.co.uk/pr-vital-to-business---watch-heathrow-ceos-prpays-interview/ [Accessed 28 January 2020].

8 Communications for building consultancies

Strategies, methods and examples

Tom Smith

Introduction

Building consultancy comprises some of the most important professions when it comes to successfully delivering complex buildings and infrastructure development. The project managers, cost consultants and quantity surveyors are involved at every stage of a project, driving quality, efficiency and sustainability, as well as maximising social, health and wellbeing elements.

Yet it has to be said that the industry perhaps doesn't get the recognition it deserves, and this can make it one of the most challenging areas in construction PR. The good news is it is not impossible to get publicity; as always, it is about finding the right angle that appeals most to the target audience.

Mott MacDonald is typical of the building consultancies referred to in this chapter. It is a 16,000-strong global engineering, management and development company. Its building consultancy business is probably the oldest part of the group and has been around for some 150 years. It comprises approximately 700 experts working on projects across the UK, Europe, Africa and Asia Pacific. It specialises in project contracting strategy, supply chain management, construction cost data, benchmarking, contract advisory services and project finance expertise in the building, transport, utility and energy sectors.

In a multidisciplinary firm like this, when the PR team is looking to get publicity or for social media content, it continually speaks to the building consultancy experts to find out what's new, exciting and challenging, with the aim of sharing the news externally, either through the press or via social media channels. The communications team is also repeatedly looking at the performance of its PR efforts, whether it is an ad hoc bit of PR or part of a larger campaign, to make sure that maximum value is gained from the content and it is being targeted at the right people with the correct message.

This chapter puts forward a framework, with examples demonstrating how this can be achieved.

Setting objectives

When beginning to plan media relations and social media activities, it is important to think about what they are intended to achieve. What are the objectives?

Are they looking to change the current status quo? Do they require people to take action? What does that action look like?

At this point, it's worth drawing the distinction between what could be seen as routine communications activities and content written for the press – a forward-looking opinion piece, for example, or a press release celebrating a project milestone – and what we would consider a 'campaign', where PR is being used as part of a longer-term strategy to change behaviours in the industry.

Just because there are four or five PR ideas to support an initiative does not make it a campaign. It is not possible to influence or change opinions with ten media outputs or pieces of content over a two- or three-week period. A campaign needs purpose and volume, but more importantly, it needs to be sustained with frequent and new content. Unless there is a regular new angle or new take on what the campaign is looking to achieve, it will fizzle out quickly.

The effectiveness of a campaign will depend largely on the timing, and ideally this is when the opportunity and appetite for change is greatest. Audiences need information to enable them to understand the importance of the cause, the specific problem and the potential solution. It needs to pass the 'so what' test, and to do this the content needs to be engaging, interesting, even shocking. The colour added will always be key to its success and its emotional impact.

However, in all cases, setting SMART (specific, measurable, achievable, relevant and time-bound) objectives or goals first will help guide targeting and the content needed to support the ambition and will ideally result in impactful ideas or opportunities for media involvement.

Knowing your audience

The first key to successfully promoting a company and the expertise of its staff is knowing the desired audience. Which audience needs to be reached? What do they care about? What influence do they have on the company? What is the message they need to be aware of and need to know? Why is it important, and what is the action to be taken?

There are usually three separate audiences for a building consultancy. The first is obviously potential and existing clients. These might include infrastructure developers, sector specific asset owners, energy and water companies, government owned companies such as Highways England or Network Rail or local authorities. Put simply, existing clients need to be reminded that they made the correct decision to appoint the building consultancy. Organisations that aren't clients need to feel that they are missing out – why doesn't their own building consultant do this, or why haven't they been advised on this topic?

The second audience is the project partners who work alongside building consultants – the architects, engineers, construction companies and anyone involved in a project development team. They might, on occasions, lead project teams, and they want to be reassured that they're working with people who know what they are doing. Good positive coverage in the media seen by these professionals can also lead to new collaborative partnerships as well.

Third, there are the experts who work in the various building consultancy fields. Publicity helps build good reputations and a desire to work for an organisation. Similarly, this third audience comprises current employees and colleagues. What encourages employees more than seeing their own company publicly recognised as playing an important part in the successful delivery of a project?

Audiences and channels of communication

Once the audience has been identified, it is necessary to understand the best ways to reach them. In other words, which publications do they read? Do they use social media? What are they sharing socially and where? Understanding this will help evaluate PR efforts and inform which content performs best on which social platforms.

The way people access or digest their news has changed dramatically in the last decade. Digital disruption delivered by 24-hour television channels, news websites and social media has led to a hunger by people to know what has happened immediately. It has also changed the ways in which many people access their news. Instead of waiting for a newspaper or a magazine, it's easier to hop onto Twitter, read a tweet and instantly know whether it is interesting enough to require further reading. Consequently, social media has become very important to companies, although many don't understand why so much time, effort and money is spent on it. For building consultancy, social media opens a door to a number of target audiences – clients, project partners, employees and potential new recruits. Followers will get to know a company and benefit from reading about its news and views on the industry and construction. Through sponsored content, people who might not know a company can be reached and informed.

As there are so many social channels on offer, time and effort should be focused on the channel that offers the biggest return on investment – in simple terms, the one that gets the most engagement.

For Mott MacDonald, that channel is LinkedIn. It offers direct access to key people in the industry – clients and fellow professionals. As the company is quite diverse and comprises engineering, management and development consultants, it offers opportunities to target specific groups of people, sectors, regions, professions and levels of seniority, so that content is not posted to all followers, all of the time.

Focusing time and energy on LinkedIn doesn't mean that other channels are ignored. Facebook, Instagram, YouTube and Twitter all are important for the building consultancy sector. It's more a question of seeing which content performs well and keeping the momentum going. This can be done by testing specific messaging.

If content doesn't perform at all, then it might not be the content's fault – it might suggest the wrong choice of channel, or insufficient organic reach. This may be the trigger to think about sponsoring content on social media, to pay to reach those people that need to be aware but perhaps don't know the company or know that it has a social media presence.

Box 8.1 How Mott MacDonald uses social media

Mott MacDonald initially established single corporate channels for the five main social media platforms – LinkedIn, Twitter, Facebook, YouTube and Instagram. Other companies took a different approach with a proliferation of regional and sector themed channels. Having a single, go-to channel makes it easier to find and follow.

Mott MacDonald celebrates its global reach, sector diversity and the positive impact it has on the lives of people all over the world. This often sees its social content relating to building consultancy, engineering and transport planning appearing alongside its work in international development. If the company had followed the same route as its competitors, then the alternative would have been separate channels for transport, energy, built environment, water and international development. In addition, there might have been channels for North America, UK and Europe, Middle East, Africa, Australia and New Zealand. The number of channel options could be endless. However, the flipside is that not all content will appeal to all followers – it's a question of finding the right balance.

Social media success is often based on volume, frequency and immediacy. It must be happening now. If content does not exist or is hard to come by, then having more than one channel will be hard to justify. A single channel is probably the best option to ensure quality.

Mott MacDonald targets its posts on LinkedIn and Facebook to ensure that the right sub-groups of followers are seeing relevant events, awards, country/region league tables or recruitment opportunities. It allows the company to create content and target specific audiences in different ways.

This sort of targeting isn't possible on Twitter using organic posts only, and the company has seen the need to create additional channels for its education business (Cambridge Education), its recruitment efforts and, more recently, its digital business. The main reason for this has been the increase in volume of posts for these areas of business. If content had been kept on the single corporate Twitter channels, the company would have tweeted at least ten times a day, which would have meant that some posts would have got lost or too little time would have been allowed for them to be digested.

Table 8.1 shows how different social media channels are utilised for posting different types of content and which are then used to re-share and amplify those posts.

Communications for building consultancies 135

Table 8.1 How content is focused and amplified at Mott MacDonald

Corporate news	Awards/Events	Interviews/features	Project insights	Recruitment	International Development	Digital content
LinkedIn • Main channel Twitter • Main channel • Share in other channels Facebook	LinkedIn • Main channel Twitter • Main channel • Share in other channels Facebook Instagram	LinkedIn • Main channel Twitter • Main channel • Share in other channels Facebook YouTube	LinkedIn • Main channel Twitter • Main channel • Share in other channels Facebook Instagram YouTube	LinkedIn • Main channel Facebook • Main channel Twitter • Recruitment channels • Share in other channels Instagram YouTube	LinkedIn • Main channel • Education channel Twitter • Main channel • Education channel Facebook • Main channel • Education channel Instagram YouTube	LinkedIn • Digital channel • Share in main channel Facebook • Digital channel • Share in main channel Twitter • Digital channel • Share in other channels Instagram YouTube

In addition to centrally controlled social profiles run by the PR team, another important way to reach new audiences online is through employee advocates. Faceless corporate social media channels might not be as trusted by social media users as perhaps they used to be. According to the Edelman 2019 Trust Barometer[1] more people trust a regular employee (53%) than a CEO (47%). Even more trust a company technical expert.

Employees are connected directly to clients, colleagues, friends and family, and they offer an authenticity and a trustworthiness that perhaps corporate channels are beginning to lose. At the same time, they might be connected to people who do not follow a company social handle.

In 2015, LinkedIn found that employees' combined followers tend to be ten times larger than the company itself.[2] This is still true of companies today. Mott MacDonald has 12,000 technical experts across all its operations. If they have an average of 500 LinkedIn connections, then together, the potential audience is six million. Obviously, there will be duplication of contacts, but the potential audience is still likely to be some 10–15 times higher than the company's current corporate page following of 350,000.

Encouraging colleagues in a business to share content online will help reach those clients, potential recruits and project partners who they are connected to on social media. It will also help to retain and engage employees, offering a potentially exciting way to get involved in social media initiatives, rather than just listening.

Podcasts should also be considered. There are now more than seven million people in the UK tuning in to podcasts, twice as many as in 2013.[3] It is therefore a platform that communications consultants cannot ignore, especially as the total number of listeners grew 24% in the last year alone.[4]

As far as publications are concerned, it is always good to rate target publications into 'tiers'.

Tier one publications are those that require regular contact – typically the construction or infrastructure trade press. There are currently only two trade publications specifically focused on building consultancy. One is *Modus* magazine, published by the Royal Institution of Chartered Surveyors, and the other is *Project* magazine from the Association for Project Management.

If the primary aim is to reach building consultancy professionals (including potentially competing consultancies), then both provide great opportunities to highlight successful projects and the roles that these experts play in overcoming construction challenges. They offer insight into best practice, advocate new thinking and new ways of working for the industry and provide opportunities for thought leadership or commentary. More importantly, they are publications that potential new recruits read and companies need to impress.

However, there are also many more publications on offer in the many sectors that building consultancy operates, providing good avenues to promote an organisation's work. In many ways, these publications offer a much better opportunity to reach clients, project partners and other external stakeholders with whom teams want to engage.

To reach a wider audience to influence decision makers and to win work, contact lists need to be broadened to include journalists in the building, construction, engineering and architecture press. Depending on the project, these can include magazines in the water, energy, healthcare, education, transportation and other sectors, and these form the tier two targets.

This is where promoting building consultancy can become quite challenging, depending on what the expectation is.

When looking to target publications such as *Construction News*, journalists will want to hear from contractors, as this is primarily their audience. *The Architects' Journal* will want to hear from architects, *New Civil Engineer* features structural or civil experts and *Property Week* wants to talk to developers and agents. Editors are unlikely to publish an article purely detailing what the building consultant did. They will, however, be interested in how, working with the other professions, building consultants supported a project's business case, overcame construction challenges, reduced spend without affecting the design or improved the project's environmental credentials with specific focus on health and wellbeing of occupants or users.

In any building consultancy, there will always be a desire for coverage in national newspapers and current affairs magazines and for spokespeople to appear on TV or radio. And while these opportunities aren't out of the realms of possibility, it takes a very compelling story to achieve national coverage. National newspaper journalists are usually not going to be interested in how a project was managed or how it was kept on time or budget. They would actually be keener to talk to you if the project was over budget and late! However, if there is an angle that is big enough and relevant enough to the public – a human angle – then it is always worth asking.

Opening up opportunities

The media landscape has changed immensely in the past 20 years. Digital disruption has changed the way many journalists work and, as a result, has nearly killed off the 24-hour news cycles of the print media. The time that journalists originally allowed to ask questions through face-to-face meetings, to research the subject and to develop articles has diminished hugely, as publications face pressure to announce news as it happens. Consequently, the trend is now for more exclusively placed news or quickly sub-edited press releases published on websites.

The digitisation of the media has also led to many publications disappearing from existence. Part of this has been caused by dwindling advertising budgets (the result of fewer subscribers) and, more recently, the impact of the COVID-19 pandemic. But it was also partly due to the inability to keep up with the current news cycles. The tendency now is for printed publications to focus less on news and more on issues and project detail, preferably as an exclusive. News is more effectively covered online, as it breaks, via websites and social media.

So, how can building consultancy be promoted in the modern era to make sure it is heard alongside the PR activities also coming from the client, architect, engineer and contractor? A lot will be down to the story pitched to publication and relationships with journalists.

Press releases remain an important mechanism. While they don't carry the same weight as they used to, with fewer journalists covering these announcements, they're still read by many and offer a searchable history of project wins or achievements on a corporate website.

As always, coverage will tend to depend primarily on three factors – how relevant, big and timely the news is, the relationship with the journalist and the reputation or brand recognition of the company. Remember, if attention isn't immediately grabbed from the headline or the first two sentences of the release, then a journalist won't read any further.

If it is possible to offer a few tier one publications sight of a press release or arrange a telephone briefing prior to the announcement date (under embargo), then it can lead to a more detailed story appearing. It will also help build a relationship with a journalist.

If a publication still maintains an editorial schedule, then forward features also provide clear opportunities to place content or to put forward spokespeople. These aren't always set in stone, and many editors or features editors will be open to ideas. Some publications will produce features themselves or commission freelance journalists to write them. However, there are some magazines that will accept project case studies from companies, as long as they're not written in an advertorial style.

OpEds (opinion pieces) require a quick turnaround if commenting on something in the news (otherwise known as piggybacking). Industry issues, controversies, new regulations and policies, all offer good options for coverage. Most magazines encourage comment submissions or letters to the editor. Once again, it comes down to the timeliness and relevance of what is being pitched. It also needs an informed opinion, of course.

What makes news for building consultancy?

A quick analysis of common coverage relating to the building consultancies sector would suggest that some parts of the media are only interested in project underperformance or cost overruns. And as this doesn't always help with the sector's reputation, PR experts need to do their utmost to talk about the positives and to extol the virtues of the various professions.

Understandably, news is very often about conflict – things going wrong, controversy, differing points of view, contractual or legal disputes. For many journalists, this is low-hanging fruit: for example, complex major infrastructure projects like Crossrail failing to open on time and £3.5bn over budget. It would be more complicated and distract from the story if the media considered that, in comparison, the Channel Tunnel went 80% over its estimated cost,[5] compared to Crossrail's current cost overrun of 23%.[6] Thankfully, once people start using a facility or a transport system, they largely forget how delayed or over-budget it was in the first place. For example, do journalists continually tell the story of how the Sydney Opera House was ten years late and 1,457% over budget?[7]

What can PR professionals do to tackle the inclination towards negativity and to get positive news coverage for building consultancies?

The answer is plenty, but only if the pitch contains a few key ingredients. First of all, to make something newsworthy, it needs to be timely or topical. News piggybacking is a quick and easy way to get coverage – as long as the story has substance and is not just parroting what has been covered elsewhere. Second, the story needs to be relevant to the publication and its readership. Third, is it unusual or new, or does it suggest changing in the way things are done? PR experts are always striving to learn about new things that could differentiate their messages from competitors and to steal a lead. And then there is the human angle – how will it affect clients, their customers, the industry or the wider public.

When building messaging, remember to include colour, detail or examples that will resonate with that audience. Have facts and figures to bring the story to life.

Building relationships with the media

When engaging with the traditional media or posting to social media, the common business objectives are usually to demonstrate success, capability and expertise, to communicate the thoughts and views of experts or to celebrate company initiatives. All of this obviously supports brand recognition, reputation, company ethics and culture, which is great.

Building consultants play a pivotal role in construction, and it could be argued they actually have the biggest impact or effect on a project's success. At pre-construction and design stages, their work decides what is possible given the budget and the required revenue, which can actually determine whether a development goes ahead and what it looks like. When it comes to the construction phase, they are the people trying to keep costs down and deliver on time.

And it is the projects that provide opportunities for publicity and association throughout the design and construction phases, allowing PR professionals to make sure messaging is tested at each milestone. It is commonplace to target the same publications as project partners, and those outlets tend to be more focused on the engineering or architectural aspects or the construction techniques. The key to success is to pick the moment or the opportunities where it is clear that there will be most return on the effort.

For example, at the beginning of a project, the appointment of the development team comprising building consultants, engineers, architects and sometimes contractors, is normally announced through a single press release produced and distributed by the client. The audience for this release is primarily the national and local media, as their readers are the people the client needs to gain approval from. The development team members are mentioned to provide reassurance that the project will be a success. To add weight to this, previous achievements may be included, assuring the public that these people know what they are doing.

Depending on the size and scale of the project, these announcements can go far and wide, which is great for brand recognition. Don't be too disheartened if coverage includes the briefest of mentions at this stage. The client obviously takes the limelight – it is their project after all. The architects might get a bit more attention, as they provide the computer-generated images visualising the project. They might also get asked by the architecture press about how the design was inspired, and they might say that the design has a connection to the local community and is sympathetic and in keeping with its surroundings. Engineers might get the opportunity to talk up the technical challenges that they will need to overcome.

There are potential angles for building consultants too, such as talking about the sustainability aspects of the project, especially if the ambition is BREEAM certification.[8] Another might be if there is a particularly challenging schedule. Both of these depend really on whether it is relevant to the publications being targeted. As always, this might be something that the client has already included in their messaging or wants to focus on, so back to the drawing board for something else!

The opportunities become more interesting as a project moves forward, and this is when, with client approval, it should be possible to talk in more detail about the role building consultancy plays.

Project milestones such as contractors being appointed, construction beginning, completion and opening or 'topping out' of certain aspects all provide another reason for a press release. And instead of working in competition with the other project communications teams, the best results are often delivered in partnership with them.

Working together allows all parties to get their voices heard. And one of the best ways of doing this through site media visits.

Participants can include the client and either the architect, engineer or contractor, depending on the publication(s) invited. There is a common protocol for these events where each company representative gives a short presentation to the media attending before going onto the site. The client and the building consultant talk about the project, what it is they're aiming to achieve, what has been achieved so far and what is coming up. They can also include any interesting aspects, such as specific new technologies being used or perhaps the social benefits that the project will provide when complete. The architect then gives insight into the design, and the engineer covers off relevant technical aspects of the design. The contractor then talks about the construction techniques that have been used and what the journalists will see when they go out on site. Lunch or refreshments afterwards is always a good idea, as these visits can take up a morning, afternoon or even the whole day.

Media site visits take time to organise, and everyone needs to make sure their messages are aligned, but they are worth the effort, as journalists are generally interested in seeing a project progress. Site visits will help build relationships with the press, and the resulting coverage tends to be more insightful.

These types of activities are essential in developing good working relationships and getting good quality coverage. Offering exclusives, whether it is a site visit, interview or briefing on a project win, will help cement the connection. However, it is also crucial to support a magazine's agenda and getting involved in their campaigns can be an easy win. As always, this will include getting involved in their commercial opportunities from time to time, but hopefully, the sponsored content will overall be less than the earned editorial.

Box 8.2 Cumbria Infrastructure Recovery Programme site visit

In 2015, Storm Desmond hit the UK hard, bringing unprecedented rainfall to the north west of England. It was a 1-in-200-year flood event, and Cumbria bore the brunt. In just 48 hours, the county was deluged with 1.15 trillion litres of rainfall, enough to fill Wembley Stadium 290 times over. Almost 8,000 homes were flooded. The storm also impacted 2,000 businesses, more than 600 farms and 300km of carriageways.[9]

Nearly 800 bridges were damaged. Several stone-built river crossings, some of which had stood for hundreds of years, were swept away. In the storm's aftermath, businesses and livelihoods were badly affected, and there was an urgent need to restore vital road links and reconnect cut-off communities to get the county moving again. Cumbria County Council initiated the Infrastructure Recovery Programme (IRP) to achieve that.

The council's primary communications audience was obviously local communities and visitors to Cumbria's famous Lake District. Communication was achieved through national and local media, the council's website and social media channels. Messaging was related to what residents needed to know to enable them to get on with their lives as best as possible.

The scale and speed of the IRP required a huge effort by multiple players, working as one team with the council at the centre. It provided a great opportunity to hit home the abundant skills, technical expertise and creativity they brought. The target audience for this was professionals working in the engineering, construction and building consultancy sectors.

Working with Cumbria County Council, Mott MacDonald organised a press visit in January 2018 with the objective of achieving three detailed articles in the engineering, construction, building consultancy trade press, while gathering content for social media and marketing use.

Journalists from *New Civil Engineer* and *Construction News* were quickly on board. As both magazines are published by EMAP, it helped offer a degree of exclusivity to two tier one publications. An in-house journalist from Mott MacDonald also attended, so that other target publications could be offered articles following the visit. A PR representative with photography skills also took part in the visit.

The collaborative culture evidenced on the project delivery was replicated in the approach to communications, making it an exceptionally rewarding media visit. With the council's IRP management team, travel to Carlisle and overnight accommodation was organised, the format and presentations for the day agreed, timing, logistics and transport to four different work sites planned and key project personnel booked to discuss the IRP's challenges and the logistical and technical solutions developed/implemented.

The day ran like clockwork: presentations in the morning at the council's offices in Carlisle, then a visit to the first site – Brougham Old Bridge near Penrith. Next was the temporary crossing at Pooley Bridge. A quick pitstop for lunch at the Boot and Shoe Inn in Greystoke was followed by a visit to the last site, Bell Bridge at Sebergham, a washup of any remaining questions and travel back at the council's offices. The level of engagement, warmth and kind welcome from the council and IRP team was fantastic. The journalists' investment of time in visiting Cumbria was rewarded beyond their expectations, resulting in strong, positive engagement with the IRP story at a human as well as technical level. It made the site visit highly memorable. It really couldn't have been done without the backing and involvement of the people at Cumbria County Council.

All the set objectives were achieved. *New Civil Engineer* and *Construction News* both ran three-page features (greater than their usual project story page allocation) within a week or two of the visit. *Construction News* featured the story and image of the day on the cover. By staggering the release of stories over the following two months, articles were placed in *Project Magazine* and *Bridge Design & Engineering* magazine.

Social media-wise, content and images were published during the visit itself and after. Publication of each of the four articles provided an opportunity to repeat messaging connected to the project.

The project is now in its fourth year and continues to offer opportunities to repurpose content. Since the visit, the project has received seven awards which have recognised the efforts of everyone involved. Mott MacDonald has also developed written and video IRP case studies. IRP was also a featured project story in Mott MacDonald's client-facing publication 'This Is How'. Using words, pictures and video, the publication shows how Mott MacDonald worked with clients and infrastructure industry partners to drive better outcomes for communities. The IRP is exemplary, as the site visit showed.

Communications for building consultancies 143

Figure 8.1 The project team answering questions from journalists on the media visit.

Figure 8.2 Rehabilitation work on Brougham Old Bridge.

144 Tom Smith

Figure 8.3 Temporary bridge at Pooley Bridge.

Figure 8.4 Team picture of the journalists and the project team.

Evaluation

Evaluating and demonstrating the impact of PR efforts and associated marketing has long been the million-dollar question. John Wanamaker, whose 16 retails stores eventually became part of Macy's in the US, is credited with coming up with the phrase, 'Half the money I spend on advertising is wasted; the trouble is I don't know which half'.[10] And to a certain extent, it is still a challenge to prove the return on investment in advertising or time spent on PR activities.

Thankfully, the international Association for Measurement and Evaluation of Communication (AMEC) has helped professionals in the public relations industry plan, action and measure how their efforts support a business strategy in a meaningful way.

The interactive evaluation framework[11] from AMEC provides a consistent and credible approach that allows PR professionals or teams to align their communications efforts with company objectives.

Planning communications can be an extremely time-consuming process. But being able to establish what the objectives are early on, what the targets are and how they will be measured, will provide rewards in the long term, helping demonstrate much more value for the PR effort as part of the business strategy. It also offers a consistent way of working with colleagues who perhaps don't understand PR or marketing, explaining where it can support their needs. Nobody should be left with the feeling that they're being asked to make time for press interviews, or help develop features or press releases without knowing what the point is or what will be achieved through it.

With more digital focused publications, websites and social media, the PR industry is now more data-rich than it was ten years ago. This makes the measurement of impact a lot more plausible, especially as it is possible to target content to certain audiences and personalise and customise messaging to each specific person. It is even feasible to customise and personalise website pages depending on who's viewing them. The data from social platforms and website analytics enable PR teams to understand more about audience behaviours, to learn what content performed well and if a call to action was received and acted upon. An example of PR impact might include: 1,000 people visited a website in one-week, 800 signed up to a webinar and 30 people got in touch asking for a meeting. In addition, there was a 200% increase in monthly coverage relating to the PR content, and a 30% increase in coverage in general.

Make sure to continually analyse and report on your PR evaluation metrics, as it shows whether your strategy is working. If not, make changes and test a bit more. And finally, when it comes to demonstrating the value of PR, it is important to publicise and shout about its success, both internally and externally. Hearing about how many people or organisations have engaged with a campaign or signed up to a commitment encourages others to join in and act. Demonstrating to colleagues internally how many people were reached with positive coverage will hopefully encourage others to seek some PR support on a project, offer input for an opinion piece, see the benefit in celebrating success and get more involved as a brand ambassador.

Notes

1. Edelman: 2109 Edelman Trust Barometer: https://www.edelman.com/sites/g/files/aatuss191/files/2019-02/2019_Edelman_Trust_Barometer_Global_Report.pdf [Accessed 28 January 2020].
2. Fast Company: Why vocal employees are a company's best PR, 25 March 2015: https://www.fastcompany.com/3044156/why-vocal-employees-are-a-companys-best-pr [Accessed 28 January 2020].
3. Ofcom: Audio on demand: the rise of podcasts, 30 September 2019: https://www.ofcom.org.uk/about-ofcom/latest/features-and-news/rise-of-podcasts [Accessed 28 January 2020].
4. Ibid.
5. *Daily Telegraph*, 25 things you might not have known about the Channel Tunnel, 6 May 2019: https://www.telegraph.co.uk/travel/destinations/europe/france/articles/channel-tunnel-facts/ [Accessed 28 January 2020].
6. *Building magazine*: Crossrail hit by further delay as cost tops £18bn, 8 November 2019: https://www.building.co.uk/news/crossrail-hit-by-further-delay-as-cost-tops-18bn/5102624.article [Accessed 28 January 2020].
7. Mashable: Building the Sydney Opera House, Chris Wild, 11 July 2015: https://mashable.com/2015/07/11/building-sydney-opera-house/?europe=true [Accessed 28 January 2020].
8. BREEAM is the BRE Environmental Assessment Method, a sustainability assessment method for masterplanning projects, infrastructure and buildings.
9. BBC News, Storm Desmond; your questions answered, 9 December 2015: https://www.bbc.co.uk/news/uk-35038617 [Accessed 8 August 2020].
10. Quotations Page, John Wanamaker (attributed), US department store merchant (1838–1922): http://www.quotationspage.com/quote/1992.html [Accessed 28 January 2020].
11. Association for Measurement, Evaluation of Communication: Integrated Evaluation Framework: https://amecorg.com/amecframework/ [Accessed 28 January 2020].

9 Communications for construction technology

Adopting a strategic approach to a rapidly changing sector

Paul Wilkinson

Introduction

Construction is one of the world's oldest human activities, evolving as people's knowledge, ingenuity and ability to use tools and raw materials improved.

The complex industry today often labelled simply as 'construction' has therefore been constantly evolving. Its industry-leading businesses are repeatedly applying new ideas and innovating to tackle new challenges, devising new construction methods, plant and materials, as clients and project teams have aimed to build faster, higher and deeper.

The first industrial revolution brought mechanical innovations, notably steam power, and started major social and economic shifts including the growth of major towns and cities and the development of highways, railways and mass transportation. During the nineteenth century, architects, engineers and other disciplines began to establish professional institutions and develop bodies of knowledge.

During the second industrial revolution (also known as the technological revolution, a phase from the late nineteenth century into the early twentieth century), the era of mass production, these same professionals applied their knowledge, and construction became increasingly regulated by national laws, contracts and regulations.

The third industrial revolution (the digital revolution) started in the 1980s and was driven by new computing and electronic technologies, with some professional tasks automated by developments in word-processing, spreadsheets and computer-aided design (CAD).

We are now on the cusp of the fourth industrial revolution, which combines hardware, software and biology (so-called cyber-physical systems) with advances in communications and connectivity.

Construction of new built assets and the operation, repair and maintenance of existing or legacy buildings and infrastructure also remains a vital economic activity, comprising, in the UK, for example, around 6% of economic output (pre-pandemic).[1] Homes, business premises, education, health and leisure facilities, utility services and transport systems all rely upon the architecture, engineering and construction (AEC)[2] or 'the built environment' industry, but while many other sectors have been dramatically transformed – even disrupted – by new technologies, AEC is only slowly embarking upon its digital transformation.

The sector remains extremely fragmented; contractors work in highly contractual, often adversarial environments with very low profit margins and poor payment practices, and many project teams are selected on their ability to deliver assets for the lowest price rather than on a best whole life value basis. As a result, information and communication technologies – the focus of this chapter – have often been regarded as overheads, not investments.

Arguably, large parts of the AEC sector are consequently still engaged in the third – digital – revolution. Before we begin to look at communication issues, it is useful, therefore, to understand the sector's recent development, its current technology status and its anticipated trajectory.

Digitising construction

The slow pace of AEC technological adoption is partly due to the necessarily conservative approach its people adopt when devising and building facilities with daily impacts on peoples' lives, health and welfare. So, while other industries have extensively digitised and, in some cases, been disrupted as a result (look at retail, banking and publishing, for example), construction has changed only slowly. Indeed, it remains almost resistant to change. A May 2019 *Construction Manager* survey found the biggest barrier to adopting new technology is 'individuals not wanting to change old processes'.[3] As a result, a 2016 analysis by the McKinsey Global Institute[4] suggested that Europe's construction industry was the least digitally adapted of all industry sectors, performing worst in terms of its volume of digital interactions and transactions.

The conservative culture is not the only factor in low levels of industry digitisation. The low-margin, cost-conscious AEC sector also invests less than most other sectors in information technologies. The Gartner consultancy's annual IT metrics show the 'construction, materials and natural resources' sector is consistently bottom of the league for IT expenditure. Against a 2017 cross-industry average of 3.5% of revenue spend on IT, construction spent 1.2% (as a proportion of operating expenses: construction spent 1.3% against a cross-industry average of 4.6%).[5]

However, the sheer scale of the AEC market – global spend is around £9 trillion (and forecast to grow to £11 trillion by 2030) – means that total annual global AEC IT expenditure is, at the time of writing, around £112 billion. A high proportion of this is on conventional IT hardware and software to support routine business office and communication functions, but there is also a considerable spend on AEC-specific hardware and software; total global expenditure on AEC software is estimated to be around €7.9 billion or £6 billion (at 2019 exchange rates), and this is forecast to grow as businesses invest in digital transformation.

BIM and wider digital transformation

For many businesses, this digital transformation has been under way for decades. For most, it probably started with the late 1980s/early 1990s computer-aided design (CAD), which digitised previously analogue, manual drawing production activities.

However, the digital momentum has grown as governments and business leaders have sought to respond to the challenges posed by the global financial crisis, climate change, population growth and increasing urbanisation. Efforts continue to address skills shortages, and to improve the sector's efficiency and productivity.

Major reports from the McKinsey Global Institute[6] and the World Economic Forum/ Boston Consulting Group,[7] among others, underline the international need to reshape construction. Individual nations, including the UK, have been taking steps to overhaul inefficient and outdated structures and practices and to embrace twenty-first-century technologies – in short, to embrace the fourth industrial revolution.

Successive UK government construction industry strategies have committed to improving project delivery times, reducing the whole life costs of construction, reducing the industry's environmental impact and making the UK AEC sector more competitive.

First published in 2013, the government's Construction 2025 targets[8] – 33% lower costs, 50% faster delivery, 50% lower greenhouse gas emissions and a 50% improvement in exports – still define its ambitions for the sector. But improvement has, sadly, sometimes been barely perceptible.

There has been a long line of industry reports on industry AEC inefficiency, including the 1994 Latham Report, the Egan Report in 1998 and Wolstenholme in 2009.[9] The most recent, Mark Farmer's 2016 Modernise or Die report,[10] highlighted how UK AEC productivity has effectively flatlined since the early 1990s. He identified under-investment in technology and innovation as one of the causes. But identification of this issue is in itself progress; highlighting the shift in less than a generation, Egan had barely mentioned IT.

Alongside profound changes to the industry's structure, processes and culture, Farmer underlined the opportunity to exploit building information modelling (BIM). BIM is a digitally-enabled collaborative process of project delivery extending through design, construction and onwards into operation and maintenance.

This is an area in which the UK has become a world leader. Since 2011, the UK government has been urging BIM adoption and set minimum BIM capability requirements for those delivering projects for central government departments to be achieved by April 2016. This was a significant commitment, as the UK government remains the AEC sector's biggest client, delivering about 40% of the industry's workload.

Strongly supported by two respected chief construction advisers, Paul Morrell and Peter Hansford (although this government post was discontinued in November 2015), a government-funded BIM Task Group helped to align government and industry activities, with several large private sector clients and major infrastructure schemes such as London's Crossrail also incorporating BIM into their ongoing project delivery processes.

The 2015 Digital Built Britain strategy[11] reinforced the direction of travel, and BIM-based working has gradually expanded.

However, as of 2020, it has yet to become the norm across the sector. After the so-called 'BIM Level 2 mandate'[12] passed in April 2016, the Task Group was eventually disbanded, and new government-funded bodies, including the Construction Innovation Hub and the Cambridge Centre for Digital Built Britain, have been tasked with maintaining the direction of travel, supported by industry-based organisations such as the UK BIM Alliance.

The future digital direction

Such developments are necessary. BIM is not the only potentially transformative technology change. Like other industries, AEC has automated office functions, started to adopt cloud-based Software-as-a-Service applications,[13] incorporated social media into its communications[14] and widely deployed smartphones and tablets.

Parallel technologies including drones, geospatial positioning tools, the 'Internet of Things', wearable devices, laser-scanning, photogrammetry, 3D printing, artificial intelligence and machine learning, augmented and virtual reality, big data, open data, 5G communications, 'digital twins', Smart Cities and blockchain now all feature in AEC technology conversations.

And again, like other industries, the AEC sector has also started to develop a heightened awareness of cyber-security, while also grappling with an industry culture of 'data silos' partially perpetuated by some major software vendors' closed proprietary data formats.

BIM and other digital technologies were and are expected to encourage new business models and new ways of delivering projects. It is no longer about digitising existing processes, but about 'digitalisation', defined by Gartner[15] as 'the use of digital technologies to change a business model and provide new revenue and value-producing opportunities: the process of moving to a digital business'.

New approaches included wider adoption of offsite manufacture – also sometimes called modern methods of construction (MMC) or design for manufacture and assembly (DfMA) – procurement on the basis of whole life value, the establishment of long-term strategic enterprise relationships,[16] and improved measurement and monitoring of whole life built asset performance – all strongly advocated in 'Construction 2025', the July 2018 Construction Sector Deal[17] and the National Infrastructure Commission's Data for the Public Good.[18]

Sadly, it has also taken a tragedy, the 2017 Grenfell Tower Fire disaster, to prompt regulatory proposals that designers, constructors and owner-operators should also maintain digital records of such buildings.

Dame Judith Hackitt called for the creation of a 'golden thread': a digital record for all new high-rise residential buildings 'from initial design intent through to construction and including any changes that occur throughout occupation'.[19] This will, if enacted, create whole life-cycle data obligations that can be scrutinised by regulators and others.

Hackitt also expressly specified that government should 'mandate a digital (by default) standard of record-keeping' covering design, construction and subsequent refurbishments. Digital records would 'be in a format which is appropriately open and non-proprietary with proportionate security controls'.

Balancing openness and security is also recognised in the Centre for Digital Built Britain's November 2018 Gemini Principles,[20] defining key aspects of government and industry thinking on 'digital twins' while also criticising the lack of integration and interoperability between software solutions due to vendors' closed proprietary data formats.

Construction technology ('Contech'): the communications universe

For some AEC businesses, therefore, digital capabilities are and will continue to be critical to their ability to deliver and operate projects. With major public sector clients and a growing proportion of private sector clients demanding BIM and other digital expertise, IT capabilities are an increasingly important differentiator for many contractors and consultants seeking to win and deliver projects and to support them through their operational lives, with tomorrow's buildings and infrastructure also set to incorporate 'smart' sensor technologies. Growing advocacy of offsite manufacturing approaches is also starting to change how projects are procured, with early involvement of manufacturers and suppliers just as likely to be recommended as early engagement with other parts of the supply chain.

The role of PR professionals

PR consultants or in-house teams working for such firms will therefore increasingly need to understand these businesses' 'Contech' capabilities[21] and the recent and future digital direction of the sector.

So, too, will PR professionals working for client organisations, central and local government departments, agencies and regulators, industry professional institutions, trade bodies, financiers, educators, insurers, lawyers and other stakeholders.

Helpfully, just as AEC industry organisations have created expert groups to focus discussion and learning, UK PR bodies have also formed special interest groups; the CIPR has a Construction and Property Special Interest Group, CAPSIG, with members drawn from a wide range of organisations,[22] as does the PRCA's Built Environment Group.[23] International Building Press[24] is another UK membership body for journalists and PR professionals working in the built environment.

PR professionals also play an important role in helping to establish and maintain the reputations of the numerous suppliers of information technology hardware and software to the construction industry.

These range from established names providing products and services to industry organisations since the late twentieth century (generic providers such as Microsoft, Apple, Google, IBM and Oracle, plus AEC-oriented software businesses such as Autodesk, Bentley and Trimble) to twenty-first-century start-ups exploiting the latest IT developments and developing new tools and applications (since the launch of smartphones and tablets in 2008, for example, numerous developers have launched mobile apps targeting just about every sector of the industry).

This, in turn, has spawned interest in 'Contech' in financial PR firms managing investor relations and merger and acquisition activities.

Media and events

AEC sector discussion of technology issues is also facilitated via media and events.

However, in common with other generic and specialist media, there has been some contraction in the range and depth of coverage devoted to the AEC market since the 1990s, with a corresponding drop in the extent and depth of conventional media coverage of IT issues.

As display and recruitment advertising revenues and print circulations and readerships have declined, some traditional print titles have ceased publication. For example, after 130 years, *Contract Journal* closed in 2009. Others have reduced their frequency or gone online-only, usually behind paywalls, sometimes under pressure from new owners. Most titles have also reduced their news, features, comment and analysis and cut their editorial teams.

Consequently, in-depth media coverage of AEC information technology issues has declined in the once dominant mainstream industry titles such as *Building*, *Construction News*, *New Civil Engineer*, *The Architects' Journal*, and *Property Week*.

On the upside, some new, online-only titles such as *Construction Enquirer* and *The Construction Index* have been launched, while the monthly *Construction Manager* magazine (associated with the Chartered Institute of Building) has an IT-dedicated website, *BIM Plus*.

However, one monthly UK title, *AEC Magazine*, continues to provide regular news and feature coverage of IT hardware and software in the sector, as well as organising a one-day technology conference called NXT BLD.

Such organisation of commercial conferences, trade shows and awards programmes has also been widely adopted by other industry publishers as well as professional events companies. Although extensively disrupted by the COVID-19 pandemic in 2020, these events are expected to be resurrected in 2021.

Large two- or three-day trade shows often feature zones for IT hardware and software businesses, with some exhibitors able to nominate speakers for parallel conferences or seminars. Relevant UK trade shows have included:

- UK Construction Week – usually in October at Birmingham's NEC
- Futurebuild – usually in March at London's ExCeL
- London Build – usually in November at London's Olympia
- GEOBIZ – usually in June at London's Business Design Centre
- Digital Construction Week – usually in October at London's ExCeL

Interspersed between these events throughout the year are a host of other events run by AEC publications. For example:

- *New Civil Engineer* organises TechFest each September.
- *Construction Manager* has organised an annual 'construction digital summit'.
- *Construction News*' Contech event was instituted in December 2019.

This role is also served by professional bodies such as the Institution of Civil Engineers, which has held BIM and digital events each autumn for several years, and the Association for Consultancy and Engineering, which organises an annual digital conference.

And, as online events have grown in popularity, many publishers and institutions have organised webinars and recorded podcasts, sometimes financially supported by industry sponsors such as technology vendors. These are expected to increase exponentially in 2020 and beyond.

Many technology vendors also run their own user conferences and exhibitions, some of them quite substantial. The annual Autodesk University[25] has traditionally attracted thousands of people to Las Vegas each November, while also spawning spin-off events in London. Bentley's Year In Infrastructure event each October also draws hundreds of attendees, for example. And Newcastle-upon-Tyne's Space Group has its two-day BIMShowLive in the city each February.

With BIM and digital working growing in importance, various special interest groups have also been established. For example:

- COMIT (Construction, Operation & Maintenance through Innovative Technology)[26] holds quarterly community days for its members, plus an annual conference.
- The UK BIM Alliance[27] holds quarterly forums, while also supporting a wide range of BIM regions and sector-specific BIM4 communities.
- Leeds Beckett University's ThinkBIM initiative[28] has a long-running series of quarterly half-day conferences.

Digital groups have also been established within many professional and trade associations, providing a focus for expert discussion and advice for members and a pool of experts to represent these bodies.

Contech and social media

The growing importance of digital technologies in the AEC sector has been paralleled by growing use of social media, particularly since the late 2000s.[29]

Hardware advances, including wider availability of broadband connectivity, shifts from analogue to digital media (for images, video, music), and, since 2008, the explosion in multi-functional smartphone and tablet usage, have put powerful digital publishing capabilities in the hands of an often highly mobile workforce.

Moreover, as the hardware became cheaper, more widely available and simpler to use, and as social platforms opened new avenues for individuals to express opinions and share content, it was rapidly adopted by often more tech-savvy, younger people. Technology conversations became more democratised – no longer contained within organisations, publications or events and less subject to 'command and control' communication strategies.

LinkedIn and Twitter, for example, became widely used by many in the construction sector, and some highly constructive [sic] conversations have taken place, some indexed by using hashtags. Since 2011 #ukbimcrew, for example, has facilitated numerous conversations about BIM in the UK,[30] while Leeds Beckett University's ThinkBIM, the UK BIM Alliance, COMIT and many other events have used Twitter, live-blogging, live video-streaming and content capture tools to share BIM and other digital events online.

Long-form content on digital issues is frequently shared on blogs and is also now a common feature of many AEC corporate websites. Wall-tie manufacturer Helifix, for example, used a blog to increase its website visibility by 45% and boosted search traffic by 40%.

Video is currently one of the fastest-growing and most effective channels used to share the latest technological developments. Established in London by Fred Mills in 2012, The B1M has become the world's largest and most subscribed-to YouTube channel for construction with over 15 million viewers per month[31] and was the first built environment channel ever to receive YouTube's Gold Creator Award when the channel surpassed one million subscribers in April 2020. Numerous AEC businesses and technology brands have commissioned videos created and distributed via The B1M.

The low-tech image of construction

While many individuals will take a keen interest in construction technologies, it is also important to remember how disparate and fragmented the UK AEC industry remains, with its strong reliance on hundreds of thousands of small and medium-sized enterprises (SMEs) and self-employed workers, particularly at the construction site level and in the repair and maintenance market. BIM and other digital developments are often a distant and irrelevant consideration to them.

The whole AEC sector is also a highly volatile sector prone to extreme economic cycles.

The 2007/8 global financial crisis pushed the UK into its deepest post-war recession, and it took ten years for construction output to return to pre-recession levels, even though the rest of the economy grew by over 10%. During the dip, tens of thousands of construction businesses, many of them SMEs, became insolvent. At the time of writing, the potential economic impact of the coronavirus crisis is expected to be much worse.

The current industry workforce is also ageing, while retiring workers are not being replaced by young industry joiners, creating what Mark Farmer described as a 'demographic time bomb'. Recruitment, he believed, is also badly impacted by public perceptions of poor job security, working conditions and health and safety, plus blacklisting, and 'cowboy builder' media exposés.[32]

The AEC sector's low levels of investment in R&D and innovation also contributes to its poor image, says Farmer, presenting a once-in-a-lifetime opportunity to change perceptions of it as a low-tech sector:

'The current pace and nature of technological change and innovation in wider society is such that unless the industry embraces this trend at scale, it will miss the greatest single opportunity to improve productivity and offset workforce shrinkage. Failing to embrace change will also further marginalise the industry by reducing its attractiveness to a new generation of workers who will have grown up in a digital world'.[33]

Twenty-first-century construction communications

Clearly, as digital transformation becomes more widespread, AEC businesses will have to (to use Mark Farmer's report title) modernise or die. And it is not just about implementing technological changes; wider cultural changes will be vital.

Envisaged changes will extend beyond processes concerned with planning, design, delivery and operation of new built assets and the equally important upkeep and improvement of existing assets.

As outlined above, Construction 2025, Digital Built Britain and other visionary documents seek a shift from narrow, often self-interested, silo-based, short-term thinking to working in more holistic, integrated and long-term relationships. This will involve profound changes.

To remain competitive, organisations will need strong and often visionary leadership. They will need to rethink some key customer, end-user and supplier relationships. How they are financed, regulated and structured may need to change. How their people are recruited, how they learn, work and develop and how they are incentivised may also need to be adapted.

Importantly, how individual organisations communicate, both internally and externally, will also need to evolve.

PR should not be regarded as a short-term tactical activity designed to present a better image but as an executive-level professional discipline whose practitioners help business leaders respond to new challenges such as digital transformation.

As this chapter has outlined, strategic communicators will need to appreciate current and future trends and so be able to help guide employers and clients towards better-informed strategic business decisions.

Improved productivity and profitability, lower waste, higher client and end-user satisfaction, better job security and increased asset quality, safety, reliability and efficiency can all be achieved through coherent investment in people, processes and technologies (in that order).

Aggregated across organisations, positive responses to such challenges and trends might also eventually change popular perceptions of the wider AEC sector. The industry's communications professionals could accumulate credible evidence, provided by data-driven measurement and evaluation systems, that enables them to collectively present a more positive image of the sector.

156 Paul Wilkinson

Achieving the Construction 2025 targets – 33% lower costs, 50% faster delivery, 50% lower greenhouse gas emissions, a 50% improvement in exports – would represent a major achievement for the AEC industry and a significant contribution to UK plc, while also helping meet global needs relating to climate change, population growth and urbanisation.

Notes

1 ONS, Coronavirus and the impact on output in the UK economy, April 2020: https://www.ons.gov.uk/economy/grossdomesticproductgdp/articles/coronavirusandtheimpactonoutputintheukeconomy/april2020 [Accessed 8 August 2020].
2 The abbreviation AEC is sometimes used to cover architecture, engineering and construction, while phrases such as the 'built environment' are also used by practitioners keen to cover the whole life of built assets; most advanced economies rely around 99% on legacy buildings and infrastructure, with repair, maintenance and facility management of existing assets more than outweighing delivery of new assets.
3 "Can new technology help us avoid low productivity?" *Construction Manager*/Kreo Software survey, 8 May 2019: http://www.bimplus.co.uk/analysis/can-new-technology-help-us-avoid-low-productivity/ [Accessed: 10 May 2019].
4 *Digital Europe: Pushing the frontier, capturing the benefits*, McKinsey Global Institute, June 2016.
5 Gartner IT Key Metrics Database, March 2017.
6 McKinsey Global Institute (2017), *Reinventing Construction: A Route to Higher Productivity*: https://www.mckinsey.com/~/media/McKinsey/Industries/Capital%20Projects%20and%20Infrastructure/Our%20Insights/Reinventing%20construction%20through%20a%20productivity%20revolution/MGI-Reinventing-construction-A-route-to-higher-productivity-Full-report.ashx [Accessed: 15 May 2019].
7 World Economic Forum/Boston Consulting Group (May 2016), *Shaping the Future of Construction: A Breakthrough in Mindset and Technology*: http://www3.weforum.org/docs/WEF_Shaping_the_Future_of_Construction_full_report__.pdf [Accessed: 15 May 2019].
8 HM Government, Construction 2025: https://assets.publishing.service.gov.uk/government/uploads/system/uploads/attachment_data/file/210099/bis-13-955-construction-2025-industrial-strategy.pdf [Accessed 8 August 2020].
9 Egan, J. (1998), *Rethinking Construction*, HMSO.
10 *The Farmer Review of the UK construction labour model: modernise or die*, October 2016.
11 HM Government (2015) *Digital Built Britain: Level 3 Building Information Modelling – Strategic Plan*.
12 The BIM Level 2 mandate required project teams to adopt a BIM process described in a series of established and prototype standards including BS 1192:2007 and the PAS 1192 series. These now form the basis of an emerging international standard, ISO 19650 (the first parts were published in early 2019), and the idea of levels has been discontinued in favour of a UK BIM Framework compliant with ISO 19650.
13 Wilkinson, P. (2005) *Construction Collaboration Technologies: The Extranet Evolution* (London, Taylor & Francis).
14 Wilkinson, P. "Application of social media in the construction industry", in Perera, S., Ingirige, B., Ruikar, K. and Obonyo, E. (eds) (2017) *Advances in Construction ICT and E-Business* (London, Routledge).
15 Gartner glossary: https://www.gartner.com/en/information-technology/glossary/digitalization [Accessed 8 August 2020].
16 Building on Infrastructure Client Group work, the ICE's Project 13 initiative http://p13o.rg.uk) [Accessed 28 January 2020] proposes wider adoption of long-term frame-

works or alliances, where integrated teams (comprising clients, funders, constructors and suppliers) work in long-term relationships to improve whole life outcomes in operation and support a more sustainable, innovative, highly skilled industry.
17 *Industrial Strategy: Construction Sector Deal*, July 2018.
18 National Infrastructure Commission (2017), *Data for the Public Good*.
19 Dame Judith Hackitt, *Building a Safer Future: independent Review of Building Regulations and Fire Safety: final report*, May 2018, chapter 8, pp.102–105.
20 *The Gemini Principles* are nine foundations for 'digital twins', aimed at delivering the National Infrastructure Commission's 2017 objective of *'Data for the public good'*.
21 'Proptech' is also increasingly widely used, but, it could be argued, the needs of property owners will also be met by deployment of 'whole life value' or 'digital twin' technologies; Contech and Proptech may converge.
22 The CIPR also has a special interest STEM group (science, technology, engineering and maths), and has produced member guidance on technology issues including open data and artificial intelligence.
23 PRCA Built Environment Group: https://www.prca.org.uk/membership/groups/sectoral/property-construction-and-infastructure [Accessed 8 August 2020].
24 International Building Press: http://www.ibp.org.uk/ [Accessed 8 August 2020].
25 Autodesk University: https://www.autodesk.com/autodesk-university/ [Accessed 8 August 2020].
26 COMIT (http://www.comit.org.uk/about-comit) was originally Construction Opportunities for Mobile IT, a government-backed research project founded in 2003.
27 UK BIM Alliance: https://www.ukbimalliance.org/ [Accessed 8 August 2020].
28 ThinkBIM: http://ckegroup.org/thinkbimblog/about/ [Accessed 8 August 2020].
29 See Wilkinson (2017), *op cit*, for a more detailed discussion of this trend.
30 Butcher, Su (2014), "The #ukbimcrew is not a clique; it's for everyone", Just Practising blog, 7 May 2014: http://www.justpractising.com/social-tools/networking/ukbimcrew-clique-everyone/ [Accessed 16 September 2019].
31 The B1M: https://www.theb1m.com/about [Accessed 8 August 2020].
32 Farmer Review (2016), *op cit*, p.40.
33 Farmer Review (2016), *op cit*, p.48.

10 Communications for sustainability

Sustainability communications in construction and the built environment

Liz Male

Introduction

In 1859, decades before the first car was patented or commercial passenger airline flight was even thought possible, an Irish scientist named John Tyndall published the hypothesis that, because carbon dioxide (CO_2) molecules can absorb heat, changes in CO_2 levels in the atmosphere would inevitably change our climate.

Carbon dioxide levels and other greenhouse gases have increased dramatically since then.[1] All the evidence shows that this increase is almost entirely due to human activity.

Consequently, the impact of climate change is a huge issue for society and a highly topical area for PR professionals, communicators and marketing teams working in construction and the built environment sectors. The UN Secretary-General has called climate change 'the defining issue of our time',[2] and climate change awareness and concerns are continuing to gain momentum around the world.

A worldwide study by Synthesio in 2019[3] analysed conversations across social networks, online communities, blogs and forums to observe people's feelings about climate change. It combed through millions of mentions of environmental issues between March-September 2019, dividing the data into three topics: Causes, Effects and Solutions. By following relevant keywords, hashtags and spikes in conversation, the researchers were also able to identify particular brands that have been enjoying success due to their stances on environmentalism.

Perhaps unsurprisingly, the research showed that a much greater number of people aged 18–25 and 25–35 are leading the online discussions surrounding human contributions to climate change compared to those in older age groups. Combined, these younger age groups account for more than six times the amount of conversations happening between any other age group online.

There is no doubt that for today's young professionals and future leaders in business, environmental topics and the impacts of climate change are significant issues and areas of interest and concern. This has become even more evident in the wake of the COVID-19 pandemic, with widespread calls to 'build back better'.

So, any construction organisation, in any part of the supply chain, would be extremely foolish to overlook this sentiment or to assume its new generation of managers do not talk openly about a 'climate crisis'.

Interestingly, the researchers saw a distinct focus in the online conversations on renewable energy as a 'fast-acting and viable, concrete solution to climate change'. Interest in solar and wind energy in particular dominated the most-used relevant keywords. This reflects the growing acceptance of the transition to clean energy – an aspect of environmental improvement in which the property and construction industries can play a huge part.

As the report concludes:

> Conversations about climate change are moving at light speed. Practically every day, global warming rattles new places, people and living things – and people talk... Social media data provides business insights on what consumers care about, which can inform a company's roadmap and investment decisions.

In 2019, the Oxford English Dictionary chose the term 'climate emergency' as its Word of the Year because of its effectiveness in communicating a sense of urgency in the fight against global warming.[4]

A national climate emergency was declared by the UK government in 2019. It followed many similar declarations around the world. As of February 2020, well over two thirds of district, county, unitary and metropolitan councils had signed up to such a declaration.[5]

Architects, engineers, building services engineers and other professionals in the built environment have also made their own climate change declarations.

From global initiatives to the hyper local level, fuelled by a wide range of advocates from individual activists to coordinated international campaigns, the drive towards increased environmental sustainability is now unstoppable, and a clear set of global and UK environmental targets are now in place that create the context for much of today's sustainability communications.

Understanding definitions

Of course, sustainability is a much broader concept than just tackling climate change (as if that wasn't big enough). Sustainability reflects a wide range of inter-related areas of risk, action and opportunity, each of which has particular importance in the built environment. There is also a significant overlap between sustainability and activities that used to be classed as 'corporate social responsibility' (CSR).

The original definition of sustainable development as set out by the Brundtland Commission in 1987 is 'development that meets the needs of the present without compromising the ability of future generations to meet their own needs'. It is often communicated using the metaphor of the three-legged stool or the concept of three pillars: economic, environmental and social, also known informally as profits, planet and people.

These interlinked aspects of sustainability are now extensive and complex, and it is easy to lose a sense of how and where an organisation can best make its impact felt. Therefore, for construction communicators and many other roles in the industry, the UK Green Building Council (UKGBC) 'Sustainability Essentials' one hour awareness online course is a very useful refresh on sustainability issues in our sector and gives us a good structure from which to start to frame our PR efforts. This course, or something very similar, should be mandatory CPD for any PR professional looking to communicate about sustainability in the built environment.

In its introductory course and in its own communications too, the UKGBC divides sustainability into five key themes:

- Climate change, including energy, carbon and net zero targets (see Box 10.1)
- Resource efficiency, including recycling and reducing waste and circular economy
- Natural capital and biodiversity
- Health and wellbeing
- Socioeconomic impact and societal change, community and people

It also highlights areas of good practice in six stages of the building lifecycle: investment, planning, design, procurement, construction and operation.

For organisations in the construction supply chain, the free training resources from the Supply Chain Sustainability School[6] are also invaluable.

Launched in 2012, the Supply Chain Sustainability School is a free learning environment, upskilling those working within, or aspiring to work within, the built environment sector. It focuses on a wide range of topics in sustainability, including:

- Air quality
- Biodiversity and ecology
- Materials
- Waste and resource efficiency
- Fairness, inclusion and respect
- Modern slavery
- Business ethics
- Sustainability strategy
- Sustainable procurement
- Environmental management
- Energy and carbon
- Water
- Social value
- Wellbeing
- Performance management

It also addresses topics in offsite construction, BIM, lean construction and management and holds CPD training and networking events across the UK.

These lists give a flavour of the wide range of areas in which it is possible to communicate today about strategy, activities and achievements in sustainability in property, construction and the built environment. The question is, which ones to choose?

Of course, different topics tend to go in and out of favour in the construction industry at different times. I noticed this particularly in the areas of interest pursued most vigorously by LMC clients since the early 2000s. For a while, it was the energy labelling of buildings that was dominant. Then natural materials. Then biodiversity, green roofs and living walls. Then health and wellbeing. Then diversity and social value.

A blog post at the end of 2019 by Robert Blood, founder of issues analysis consultancy SIGWATCH,[7] highlighted five sustainability trends for brands to watch in 2020. These included veganism and clean beauty, but also three areas which are particularly relevant to construction: the ongoing pressure to reduce use of plastics, concerns about deforestation (with more demand for companies to provide third party proof of thorough audits and certification for timber products, for example) and the need to demonstrate a commitment to low carbon impacts.

The truth is that all of these topics, and many more, should be talked about all of the time. So it is now perfectly correct for construction PR professionals to address all the sustainability themes listed above.

In each of these areas lies enormous opportunities for organisations to make progress in sustainability and opportunities for PR professionals to communicate key achievements and lessons learned. The media and public's appetite for insight across all themes is not likely to fade away.

If you're seeking inspiration from how others have done it, the UKGBC regularly highlights such examples of leadership, innovation, collaboration, good governance, commitment and performance and transparency and openness by organisations throughout the built environment. It publishes an annual 'Leading the Way' report with a full sustainability review of its Gold Leaf members across the five key impact areas, reviewing and analysing their commitments and collecting best practice examples.[8]

Major industry events such as UK Construction Week and Futurebuild also provide plenty of inspiring content, and it will be fascinating to note how much the discussion on environmental sustainability and resilience will have developed post-pandemic, once these events return in 2021.

For film and videos that communicate on sustainability, check out the EVCOM Clarion Awards. The Business in the Community Responsible Business Awards also publicise businesses that are making a significant impact in eight responsible business areas, including environmental leadership, equality and connected places, while the Ashden Awards recognise and reward innovation and sustainable energy solutions that cover everything from fuel poverty to clean, affordable electricity that enhances global healthcare, education and employment.

162 Liz Male

Box 10.1 Communicating about reducing carbon emissions

The Greenhouse Gas Protocol[9] (GHG Protocol) is a multi-stakeholder partnership of businesses, non-governmental organisations, governments and other organisations formed in 1998. It aims to develop internationally-accepted greenhouse gas accounting and reporting standards and tools and to promote their adoption in order to achieve a low emissions economy worldwide. Use of such reporting tools is very helpful for construction communicators.

It is also important for any organisation claiming to be reducing its carbon emissions to be clear about where these reductions are coming from. Therefore, carbon emissions are grouped into three categories by the GHG Protocol:

- Scope 1 – All direct emissions from the activities of an organisation or under its control, including fuel combustion on site such as gas boilers, company cars and fleet vehicles and air-conditioning leaks.
- Scope 2 – Indirect emissions from purchased or acquired electricity, steam, heat, and cooling.
- Scope 3 – All other indirect emissions arising from the activities of the organisation, occurring from sources that they do not own or control. These are usually the greatest share of the carbon footprint of an organisation, covering emissions associated with its supply chain, business travel, procurement, waste and water.

Scope 3 emissions are often the biggest and by far the hardest to measure and to tackle. Although there is now a GHG Protocol standard for Scope 3 – the Corporate Value Chain (Scope 3) Accounting and Reporting Standard[10] – it is not used much in UK construction yet.

To help with this, new guidance from the UKGBC in 2020 has been specifically developed to promote common approaches to reporting Scope 3 emissions.[11] It aims to provide clarity on interpreting the GHG Protocol standard for commercial real estate companies and to enable consistency in reporting across the sector.

Such guidance helps PR professionals to ensure accuracy. But good practice in communications also encourages PR professionals to put key figures into context, especially when communicating in terms of 'tonnes of CO_2'. This is why emissions are so often referenced in terms of car journeys, flights, balloons or other relatable items in attempts to aid public understanding. So, a carbon footprint of about ten tonnes a year is equivalent to filling 24 million balloons with carbon dioxide gas, for example.

When it comes to explaining a person or household's overall 'carbon footprint', the most common metric is CO_2eq, short for carbon dioxide equivalent. This covers all greenhouse gas emissions that contribute to climate change, including carbon dioxide (CO_2), methane (CH_4), nitrous oxide (N_2O) and refrigerant gases like hydrofluorocarbons (HFCs). The average consumption footprint of the average person living in the UK in 2018 was 13 tonnes CO_2eq per year.[12]

Heating the average UK home produces 2.34 tonnes of CO_2eq annually, according to data from the Committee on Climate Change, and a passenger's carbon footprint for a return flight from London to Malaga is 320kg CO_2eq, based on figures from the Carbon Neutral calculator.[13]

The race to net zero

The UN climate change conference in 2015 (also known as COP 21 – see also Box 10.2) led to the Paris Agreement, which in turn led to the UK becoming the first major economy in the world to pass laws to end its contribution to global warming by 2050.

The Paris Agreement aims to stimulate action to keep the global temperature rise well below 2 degrees Celsius above pre-industrial levels and to pursue efforts to limit the temperature increase even further to 1.5 degrees Celsius.

In support of this, the UK now aims to bring all greenhouse gas emissions to 'net zero' by 2050, compared with the previous target in the Climate Change Act of at least 80% reduction from 1990 levels. That now means 100% of the emissions from all our buildings, transport, farming and industry will have to be avoided completely or offset by planting thousands of trees or via other carbon capture technologies.

Such deep cuts in emissions are considered by some as just about achievable with known technologies and within expected economic costs, but these ambitious targets do also require strong and swift policy changes and extensive behaviour change.

The net zero by 2050 target has been adopted by the political administrations in England and Wales, while Scotland has committed to reducing greenhouse gas emissions to net zero by 2045, five years ahead of the rest of the UK. At the time of writing, a Northern Ireland commitment is yet to be ratified.

Since this announcement by Westminster in June 2019,[14] the rush to claim 'net zero' firsts has been particularly noticeable, including in construction and the built environment. However, some of these claims lack technical rigour and proof, most likely out of simple confusion about what 'net zero' really means.

Consequently, one of the most important pieces of work in construction and property sectors recently has also been the publication of the UKGBC's framework definition for net zero carbon buildings.

As part of a global attempt by the World Green Building Council to get every country-level GBC to develop its own campaign in a way that suits its local stakeholders, UKGBC started this work in October 2018. Published at the start of 2020, the UK framework now sets out what a genuinely net zero carbon building looks like, in two key respects:

- Net zero in operation – for existing commercial real estate, including whole buildings or even parts of buildings such as a tenanted area in a multi-let building.
- Net zero in construction – relating to new build or major refurbishment projects, where the embodied and operational emissions up to the end of construction stage are reported and offset.

Crucially, the UKGBC framework definition makes clear that the ultimate aim is to go further than this and to be able to measure the whole life emissions of our built environment:

'Many focus purely on their direct emissions rather than those that happen further along their supply chain. For developers in particular, the latter can make up more than 90% of their carbon footprint, so excluding them from scope seems rather meaningless in the grand scale of the climate crisis we face'.

This direction of travel is expected to drive the industry much more towards understanding the importance of the circular economy. But this is not yet easily measurable, so it is expected that an additional 'net zero for whole life' definition will be added to the framework within the next five years.

Box 10.2 UN Global Goals for Sustainable Development

Another important output from COP 21 was the UN's Sustainable Development Goals (SDGs) – a series of 17 inter-related and interdependent targets for humankind in 2030, with 169 specific targets sitting behind them.

The SDGs are 'the blueprint to achieve a better and more sustainable future for all. They address the global challenges we face, including those related to poverty, inequality, climate change, environmental degradation, peace and justice'.[15]

The goals provide a very useful framework for construction communicators too, allowing us to benchmark our organisations and clients against each goal and to facilitate a discussion about how much more we can do to meet each one.

The Sustainable Development Goals are:

1. No poverty
2. Zero hunger

3. Good health and well-being
4. Quality education
5. Gender equality
6. Clean water and sanitation
7. Affordable and clean energy
8. Decent work and economic growth
9. Industry, innovation and infrastructure
10. Reducing inequality
11. Sustainable cities and communities
12. Responsible consumption and production
13. Climate action
14. Life with water
15. Life on land
16. Peace, justice and strong institutions
17. Partnerships for the goals

An interesting example of how the goals are used in a construction organisation's marketing and communications is Bioregional, the charity and social enterprise behind the One Planet Living framework.[16] Bioregional has put the SDGs at the heart of its marketing campaign for some years and uses them very successfully at shows such as Futurebuild to engage with other professionals in the construction industry. Bioregional has also created a very useful practical guide which explains why the goals are so important and how the sector is uniquely positioned to help achieve them.[17]

Other businesses that have built a reputation for alignment with the goals include Cundell, Royal BAM Group, Ilke Homes, Mott MacDonald, Greencore Construction and many others.

Barratt Developments plc set out to align its sustainability strategy to the goals, commissioning an expert to analyse the meaning and relevance of each before starting on a staged plan of implementation.

According to the company's own sustainability reports, this involved:

- An initial sense check to discount SDGs that were important but nonetheless fell outside Barratt's primary business focus e.g. zero hunger.
- A full analysis of the remaining SDGs to determine their relevance to Barratt's geographical context (the UK), its industry (housebuilding), its businesses principles and priorities and the material issues determined in its 2019 materiality research. From this analysis, eight SDGs were shortlisted for consideration.
- Consultation with key internal stakeholders to verify the shortlist chosen and select the final SDGs for adoption based on where Barratt can have the biggest tangible impact.

- A review of the chosen SDGs and the selection process by Barratt senior management.

'Practically, this means the business will focus on aligning existing initiatives (and developing new ones) to deliver against these goals. It also means we will report transparently on our progress against them', says the company:

> We understand that a critical element of delivery towards the selected SDGs will be reporting openly not simply on the successes but also the areas for further improvement. This will help our business move forward with the innovation and change required to deliver these goals and to make a meaningful contribution locally and nationally.[18]

Science and the need for technical understanding

As must already be clear in this chapter, even given the very high level summary of sustainability definitions and issues so far, effective sustainability communications in construction is much deeper than it may look at first.

It demands a fair amount of technical knowledge and understanding. It is also very fast moving, with new political, scientific, social and economic developments happening day by day.

This requires PR professionals to stay abreast of the issues and to work in close collaboration with sustainability managers and technical experts in their organisations.

And because so much of a PR professional's role is often to help communicate technically complex or subtly nuanced aspects of competitive difference to largely non-technical audiences, it is helpful for PR professionals to understand the principles around the effective communication of science and how to increase public understanding of science – itself a discipline that has significantly grown in sophistication over the last 30 years.

Communicating scientific issues, including many aspects of sustainability in the built environment, has to be done with accuracy, precision, clarity, creativity and context.

The number one rule of good communication – know your audience – is vital here, and this is where the advice and skills of PR professionals in the sector can really maximise the benefits of an organisation's sustainability initiatives and achievements.

By understanding the right way to talk about sustainability (avoiding the simplistic 'eco', 'green', 'fair', 'natural' or 'environmentally friendly' shorthand) and by building genuine relationships based on mutual understanding, many target audiences and stakeholder groups can get huge value from sustainability communications and will appreciate the work of the construction supply chain much more.

As J. Gregory noted in a 2001 paper on the public understanding of science,[19] over recent years, there has been a fundamental shift in understanding how best to build public appreciation of science (with the emphasis on appreciation, not just knowledge):

> Instead of peddling factual prescriptions, scientists should work with the particular problems and expertises of the people, and tailor their advice accordingly – the task is less one of propaganda and more one of negotiation. This is clearly a more difficult task, but it is one that allows scientists and the public to work together as citizens of a scientific culture.

In many respects, exactly the same applies to architects, building services engineers, surveyors, environmental consultants and the many other industry specialists who want to communicate their messages on how buildings should be designed, constructed, used and maintained in order to reduce carbon, for example.

Collaboration, context and building trust is key, underpinning reputation, influence, relationships and mutual understanding – everything that PR is about.

Tackling greenwash

Unfortunately, 'greenwash' is probably the single biggest threat to building trust and confidence in environmental initiatives, and it is far too prevalent in the construction sector. Greenwashing undermines any organisation's genuine journey to improving sustainability.

Like fake news, greenwashing creates confusion, misunderstanding, distrust and cynicism.

'Greenwashing' is defined in Wikipedia as 'a form of marketing spin in which green PR (green values) and green marketing are deceptively used to persuade the public that an organisation's products, aims and policies are environmentally friendly and therefore "better"'.[20]

In the early 2000s, greenwashing was very widespread, including in construction. It soon became very clear that many of the 'greenest' innovative buildings of the era were found to be seriously underperforming and failing to live up to their green claims. Many building products, technologies and systems were equally put under the spotlight and, despite their 'eco' branding and product marketing laden with aspirational images of trees, flowers, blue skies and smiling children, found to be touting very dubious sustainability credentials.

A useful guide to greenwashing was published by a marketing firm in Canada. TerraChoice (since acquired by UL) conducted a study into environmental claims in the North American markets and published 'Sins of Greenwashing'[21] studies (see Box 10.3).

In 2007, the study found that sins one and two were by far the most common errors made in green marketing campaigns, both in consumer and B2B markets. It was certainly the same in the UK, typically shown in the way that completely unsubstantiated claims were being made in brands' marketing and/or the focus was put onto promoting just one aspect of environmental performance while glossing over anything that might be less favourable.

Box 10.3 Seven sins of greenwashing

1. **Sin of the hidden trade-off**
   A claim suggesting that a product is green based on a narrow set of attributes without attention to other important environmental issues. Paper, for example, is not necessarily environmentally preferable because it comes from a sustainably harvested forest. Other important environmental issues in the paper-making process, such as greenhouse gas emissions or chlorine use in bleaching, may be equally important.

2. **Sin of no proof**
   An environmental claim not substantiated by easily accessible supporting information or by a reliable third-party certification. Common examples are facial tissues or toilet tissue products that claim various percentages of post-consumer recycled content without providing evidence.

3. **Sin of vagueness**
   A claim that is so poorly defined or broad that its real meaning is likely to be misunderstood by the consumer. All-natural is an example. Arsenic, uranium, mercury and formaldehyde are all naturally occurring and poisonous. All-natural isn't necessarily green.

4. **Sin of worshipping false labels**
   A product that, through either words or images, gives the impression of third-party endorsement where no such endorsement exists; fake labels, in other words.

5. **Sin of irrelevance**
   An environmental claim that may be truthful but is unimportant or unhelpful for consumers seeking environmentally preferable products. CFC-free is a common example, since it is a frequent claim despite the fact that CFCs (chlorofluorocarbons) are banned under the Montreal Protocol.

6. **Sin of lesser of two evils**
   A claim that may be true within the product category but that risks distracting the consumer from the greater environmental impacts of the category as a whole. Organic cigarettes or fuel-efficient sport-utility vehicles could be examples of this sin.

7. **Sin of fibbing**
   Environmental claims that are simply false.

Communications for sustainability 169

There is now plenty of long-established guidance on how to avoid greenwashing.

About 20 years ago, Defra (the UK Government's Department for Environment, Food and Rural Affairs) first published its Green Claims Code and then its Green Claims Guidance to provide advice to business for clear, accurate, relevant and substantiated environmental claims on products and services or in marketing and advertising. That guidance was last updated in December 2016[22] (see Box 10.4).

The Advertising Standards Authority (ASA) also has two codes of practice (one for broadcast advertising, the other for non-broadcasting advertising) which have specific requirements on environmental claims.[23]

> **Box 10.4 The UK government's advice on green claims**
>
> In response to concerns about green marketing claims that exploded in the early 2000s, guidance was published by Defra and by other regulatory and enforcement bodies like the Advertising Standards Authority (ASA).
>
> The current Defra guidance states the following.
>
> **Principles of making an environmental claim**
>
> When you make an environmental claim for your product, service or organisation, you should make sure:
>
> - It's not misleading
> - Your messages are clear and accurate
> - The data you use is objective and transparent
>
> **Make sure your claim isn't misleading**
>
> Make sure that what you're claiming:
>
> - Doesn't cause another environmental problem
> - Doesn't suggest a greater benefit than it does
> - Isn't claiming a benefit from something in common use or from the absence of something not in common use (for instance '100% CFC-free aerosol')
> - Is something you've chosen to do beyond the legal requirements for your product
>
> **Make fair comparisons with competitors**
>
> When making comparisons, you should only compare your process or product with:

- Your previous process or product
- Another organisation's process or product
- An industry standard

If you compare your product or process with a competitor's, you should make sure your claim:

- Uses standard measurements, like miles per gallon (mpg) for vehicle fuel consumption
- Compares only with direct competitors that do the same job in the same category and marketplace
- Doesn't suggest an advantage over a rival product if there is no information to support the comparison

Use relevant pictures

You should only use pictures that relate directly to the benefit you're claiming and don't suggest something more.

You should only use logos or symbols that represent the environmental standards your product has been officially certified for.

Make your messages clear and accurate

Your claim should:

- Accurately represent the scale of the environmental benefit your product provides
- Describe a specific measurable impact or process – don't use vague terms like 'eco' and 'environmentally friendly'
- Use specific measurements or standards, for example, ISO, CEN, BSI
- Use plain, specific language without jargon
- Be clear if it refers to the whole organisation or just one area of your business

Only use symbols, pictures or labels that directly support the claim you're making.

Data to support your claims

Make sure any data you use is:

- Factual and referenced – think how a consumer might interpret your claim
- Agreed – don't base a claim on something that's not generally agreed by scientists
- Current – use the latest guidance, methods and measures in your claim

- Clear about the approach you've taken
- Available to anyone who wants it
- Reviewed and updated over time to keep it relevant

If your claims are about a target you're aiming to meet in the future, make sure:

- Details of any claims are publicly available
- It's feasible you could achieve the aim
- You make information about your progress publicly available

It is important to note that Defra has no enforcement role on environmental claims except for the EU Ecolabel (a mark that shows if a product or service meets a specific Europe-wide environmental standard and which is likely to fall out of its regulatory control post-Brexit anyway).

Instead, any policing of green claims tends to be largely left to the market itself, with competing organisations or campaign groups reporting 'unjustified' marketing claims to the ASA or to Trading Standards.

There is also a raft of consumer and business protection regulations which require certain products' environmental information to be disclosed or which aim to tackle misleading marketing claims. These laws include the Consumer Protection from Unfair Trading Regulations, the Business Protection from Misleading Marketing Regulations and the long-established Misrepresentation Act.

The basic rule of law is that you must describe your product accurately. This means that if you make a claim about your product, you must be able to prove what you say. The same principles are embedded into the CIPR Code of Conduct for PR professionals. Credible sustainability communications in construction and the built environment, as in any sector, demands total transparency, openness and honesty.

Greenblushing

While greenwashing is clearly damaging to corporate reputation and to society's progress towards sustainability, so is the opposite extreme.

According to UL,

> Greenblushing is the opposite of greenwashing; instead of providing buyers with misleading information about their sustainability efforts, companies that engage in greenblushing disseminate little or no information about their social and environmental sustainability practices or the environmentally-positive characteristics of their products.[24]

In fact, it was a PR firm in the US, Dix & Eaton, which first coined the term. It described the symptoms as:

- Believing you need all the answers before you can communicate
- Reluctance to talk about your activities, even when asked to do so
- Downplaying your achievements internally (which can have a demotivational effect on employees)
- Being afraid to discuss your efforts with customers for fear of backlash if they don't share your interests
- Always assuming there's more risk than reward in communicating
- Feeling that what you're doing is 'not that special'[25]

Of course, nobody wants to be called a hypocrite. So it's easy to see why many PR professionals in the built environment sector, who might otherwise have considered putting environmental performance at the heart of a PR campaign, could be put off by the ease by which a company can be accused of greenwashing and by the amount of work needed to demonstrate strong green credentials.

But there is now plenty of advice available and examples of good sustainability communications which can make that job easier for the PR team. And the benefits of responsible PR in this area are significant, not least in helping an organisation win business from private and public sector markets which routinely screen suppliers for proof of environmental policies and performance.

The rigour required to gain trust for a brand's integrity in sustainability also has knock-on benefits for the way that it communicates in a host of other areas.

Thankfully, some of the most vocal critics of greenwash are now much more likely to want to help. 'The environmental NGOs who used to attack companies for making exaggerated green claims are increasingly willing to work with firms to improve performance and promote genuine green strategies', noted *BusinessGreen* in 2011.

The business case for positive, proactive sustainability communications is sound. What matters now is how we go about it.

Third-party certification

One way to overcome a lot of the hurdles of deciding what is safe, accurate and verifiable when it comes to sustainability in the built environment is to seek independent, third-party certification for key achievements.

Ideally, such an approach starts very early in an organisation's thinking and planning. In this respect, PR professionals in construction can encourage collaborations and ways of working with environmental NGOs and auditing organisations who can help firms to improve performance and promote genuine green strategies.

Some major developers have embraced this approach by working with expert partners to set 'science-based targets', to work towards decarbonisation and to stay in line with the Paris Agreement trajectory.

The first commercial real estate company to do so was Landsec, the largest commercial property company in the UK, working with the Carbon Trust. Its targets were approved by the Science Based Targets initiative in December 2016, a move which 'undoubtedly enhanced our reputation and relationship with investors', according to Landsec's energy manager:[26]

> Ultimately, the science brings meaning, and grounds our ambition in reality: targets are no longer numbers pulled from thin air, they are goals linked to a real issue. Science-based targets commit us to what is required, not just what is achievable. In this sense, they prove leadership and provide the 'spine' of a long-term sustainability strategy.

More recently, in February 2020, Barratt Homes also announced its own science-based targets, becoming the first major housebuilder to take this approach. Barratt has committed to reduce its direct carbon emissions from its business operations (such as offices, sites and show homes) by 29% by 2025. It will cut its indirect carbon emissions (such as those coming from its homes over their lifetime and from across its supply chain) by 11% by 2030.

These companies are now committed to regular reporting on progress against these targets.

An interesting tip on internal communications and management buy-in is raised by Landsec's energy manager:

> Another challenge was how to make the link between the macro issue of climate change, which people see on the news, and the specific details of a science-based target. In this sense, the internal consultations and workshops were really important. We started with the sustainability team and moved out, via more senior directors who we knew were interested in these issues (the 'early adopters'), to the most senior reps who we needed to convince. By having others on board already, and by being able to show how the science informs the target and links back to the global situation, it was much easier to get sign off from the top. We had a really powerful message that empowered people and made the ambitious targets much more palatable.

Green certification schemes

Another popular option for organisations wanting to prove their green credentials, and therefore the source of many a PR story, is to get third-party certifications or ratings for particular projects or even individual buildings. Indeed, many local authorities now require some sort of sustainability rating for newbuild schemes as a condition of planning.

Examples of rating schemes and sustainable construction certification schemes currently referenced most often in the built environment include:

- Energy Performance Certificate (EPC) – a basic assessment of the potential energy efficiency of a building and a legal requirement for most buildings for more than ten years. Ratings for buildings are given on an A–G scale, with A being the most energy-efficient – and G being the least. The average EPC

rating for a home in the UK is D, but newbuild properties would be expected to score much higher.
- BREEAM – the BRE's Environmental Assessment Method which covers a number of lifecycle stages such as new construction, refurbishment and fit out, and occupied buildings, plus infrastructure (the former CEEQUAL scheme) and other sectors in the built environment. Ratings start at 'Pass', then go to Good, Very Good, Excellent and Outstanding. The scheme is very widely used and referenced by construction clients, developers and planning authorities. For example, in central London, new developments must achieve at least a BREEAM 'Excellent' rating in order to be considered for planning approval.
- LEED – the Leadership in Energy and Environmental Design scheme came from the US Green Building Council and is now the most widely used green building rating system in the world. Ratings start with 'LEED certified project' and go up to Silver, Gold and Platinum levels.
- The WELL Building Standard – launched in 2014, this is a scheme that focuses on human wellbeing: in its own words, 'the premier standard for buildings, interior spaces and communities seeking to implement, validate and measure features that support and advance human health and wellness.'. It can be used across residential and commercial projects, newbuild or existing buildings and awards ratings of Silver, Gold or Platinum.
- SKA – this is an environmental assessment tool for interiors operated by the RICS, designed specifically for office, retail and higher education fit-outs. The final rating is classified as Bronze, Silver or Gold.
- Living Building Challenge – a North American green building certification programme applicable to landscape and infrastructure projects, renovations and new buildings. It is just starting to get interest in the UK. There are two forms of certification: the Living Building Challenge Certificate or the Petal Certificate.
- Fitwel – another international standard and metric starting to gain interest in the UK. Like WELL, it is a certification system for optimising building design and operations to support human health and wellbeing. It was originally created by the US Centers for Disease Control and Prevention and US General Services Administration. Fitwel Champions are committed to implementing the standard across their real estate portfolio. Fitwel provides certification pathways for new construction and existing buildings.
- One Planet Living – a framework developed by Bioregional which looks at ten aspects of social, environmental and economic sustainability. Developers can work with Bioregional to agree and implement an action plan and achieve Global Leader status in One Planet Living.

These rating systems (and many others now in use in the UK) all have different objectives, with some placing more emphasis on environmental matters and others more interested in the health and wellbeing of building users.

Communications for sustainability

They are not without their drawbacks too. Ratings are often given at a particular point of time, may not reflect the real in-use performance of buildings or may become out-of-date quite quickly as a building becomes less efficient. Labels and certifications must be chosen carefully to be relevant to the market and to key stakeholders. Too many labels also create 'label fatigue' and even cynicism in target audiences, not least the media.

That said, the general view is that certified buildings tend to better all-rounders. Research by Green Street Advisors confirms this perception[27] and notes that there is a material 'brown discount' on rental and sales values for buildings with poor environmental credentials which tend to suffer from higher vacancy rates and higher costs.

For building product manufacturers, Environmental Product Declarations (EPDs) are particularly useful, providing a list of contents and their environmental impacts structured around lifecycle assessments (typically 'cradle to gate' or 'cradle to grave'). However, EPDs are factual and non-judgemental, so are not themselves a 'green certificate'. They are usually valid for five years.

The same applies to the British Standards Institute's suite of environmental standards (ISO 14000 series), including ISO 14001 on environmental management and ISO 14063:2010 which is the current British and European standard covering environmental communication.

Effective messaging for sustainable living

As policy makers and climate change campaigners regularly tell us, it's not enough just for businesses and organisations to change their ways in order to reach our ambitious targets on sustainability and then bask in the warm glow of positive publicity for such achievements. It also requires rapid, widespread public behaviour change.

This is why all the parties involved in planning, designing, constructing, operating, maintaining and refurbishing the UK's most green buildings also need to be supported by effective B2C communication and public education if they are to achieve the sort of sustainability transformation to which they aspire.

In this way, sustainability communications is not just PR with a green sheen. It's actually part of the sustainability solution itself.

A classic example of this is in public consultation for new development and infrastructure (addressed in other chapters in this book). Effective communications at this stage have a very direct and immediate impact on the sustainability outcomes of a project, ensuring that the needs of current and future generations can be fully built into the planning process and environmental impacts are appropriately avoided, managed or mitigated.

Another example would be the role of PR in building greater social capital. Typically, this might involve using thought leadership and communication campaigns that help to create new or stronger communities – communities of interest, or behaviours, or communities built around a place, maybe. Often, this can help facilitate significant change by establishing new social norms and can also build resilience and sustainable growth within those communities.

It is also worth considering what sort of sustainability messaging has most impact on people's emotions and behaviours.

For example, faced with constant reminders of 'catastrophic climate change', 'widespread environmental degradation', 'melting ice caps', 'species loss', 'resource scarcity', 'toxic air quality', 'societal breakdown' and even 'human extinction', are we more or less inclined to make the changes to our behaviour that everyone says are necessary?

An excellent treatise on this topic is in Futerra's 'Sell the Sizzle' report.[28] This report points out that negative messaging is actually very counter-productive and even prevents mass behaviour change, while a much more positive narrative about what a low carbon future could look like encourages greater engagement and emotion around hope, a sense of progress and excitement about tomorrow. By building a new 'availability heuristic' with positive images of the sustainable lifestyles we want to promote, we have a much greater chance of achieving them.

As the report says, 'climate change is no longer a scientist's problem – it's now a salesman's problem…. We must build a visual and compelling vision of low carbon heaven.'

Futerra's blueprint sets out advice at each stage: Vision, Choice, Plan, Action. It encourages communication that is more visual, more local and relevant, using real people and real projects to tell stories that resonate. 'Leave the hand wringing to the activist and get on with promoting your solutions', it urges.

The importance of effective imagery

As Futerra's work emphasises, the success of sustainability communication is also highly influenced by the imagery that is chosen.

Environmental researcher Gavin Lamb writes this for the UX Collective website:

> Hundreds of thousands of images and videos of climate change now circulate online. Yet with this flood of climate imagery, it's becoming apparent that we need to tell new visual stories about the causes, impacts and solutions to climate change now more than ever. We need to move beyond our standard visual vocabulary of billowing smokestacks, people admiring solar panels, and polar bears adrift on melting ice caps to represent climate change threats and solutions.[29]

He outlines his 'eight principles for impactful communication of climate change', encouraging designers and communicators to move away from the use of clichés or persistently negative and distressing images that simply try to get attention through shock tactics. Using the latest research into the effect of evidence-based climate visuals, published by an interdisciplinary team at non-profit photo agency Climate Visuals,[30] his recommendations are:

1. Hook your audience – using imagery that surprises and captures attention early on.
2. Show real people, not staged photo-ops – ideally, images with one or two human or animal subjects in naturally occurring circumstances (and avoiding politicians in the picture, in order to enhance trust in the credibility and authenticity of the image).
3. Tell new stories – exploring more unconventional imagery, showing solutions in action. Show real work happening rather than technology 'still life' pictures.
4. Show climate causes and behaviours at scale so that people get a true sense of the impact.
5. But alongside these emotionally intense climate impact images, create clear calls to action that guide people on what they can do to help.
6. Show local (but serious) climate impacts – for example, local scenes of flooding or air pollution that resonate more strongly with local residents.
7. Be extra careful with protest imagery, which tends to generate mostly cynicism.
8. Understand your audience (and their politics) – a fundamental principle of good PR, in which we research our target audiences fully to understand what will be most persuasive and meaningful to them.

PR professions in the built environment are well placed to commission photography that documents the industry's positive actions and commitment to tackling climate change and increasing environmental, social and economic sustainability. This means imagery of people as much as it does buildings or infrastructure. There is no reason to be dependent on overused, clichéd visuals.

Sustainability and the PESO model of PR

As the CIPR skills guide on Corporate Responsibility says:[31]

The activity that a PR professional might be asked to communicate under the CSR banner is powerful stuff. A client might have received an award for having policies that encourage and support a diverse workforce; they might be involved in recruiting ex-offenders or young people from the care system; a client might have made their business carbon neutral or have set up a long-term, sustainable partnership with a charitable partner.

If this activity has been done by a business because they believe it is the right thing to do as a responsible business with clear alignment to their core business operations, then a PR professional should be confident about shouting about it. If it is an ill thought out, half-hearted attempt to be seen to be doing the right thing, then it will neither generate value for money nor raise profile for a business.

This is an important reminder of PR's fundamental role as a strategic management discipline.

Sustainability communications activity should be directly linked back to an organisation or project's specific objectives, targets and strategy. And if we see

178 Liz Male

the opposite happening, we have a duty to call it out. Unsustainable practices lead to real damage to people and the planet and increasingly mean serious reputational risk.

Thereafter, throughout all aspects of sustainability communications, the basic principles of effective PR apply.

This includes identifying SMART (specific, measurable, achievable, relevant and time-bound) business and communications objectives, researching and genuinely understanding the target audience's pain points, needs and interests, creating compelling stories about features and benefits around which you can engage with your publics, delivering easy-to-understand proof points, testimonials, case studies and other content through the media and other channels and measuring impact and effectiveness to identify how to keep improving your communications.

As the ISO 14063 introduction states, 'The most effective environmental communication process involves ongoing contact by the organisation with internal and external interested parties, as part of the organisation's overall communications strategy'.

'It is time to shelve the immature and fluffy green communications and PR campaigns that have created plenty of buzz and not much else, and adopt the hard-nosed transactional-focused business-to-business marketing techniques that have proven so effective for other emerging sectors', said James Murray, editor of *BusinessGreen*, some years ago.[32] This is a message that the sector's best PR professionals have taken to heart.

So ultimately, what does a sustainable built environment really look like? To paraphrase the UKGBC again:

- It uses little or no energy, with zero greenhouse gas emissions.
- It is decarbonised and uses renewable and low carbon energy supplies.
- It uses responsively sourced, ideally renewable, materials.
- It generates no waste and facilitates a circular economy.
- It enables resilience and adaptability to cope with climate change impacts.
- It supports nature and actively enhances biodiversity.
- It is designed with the end-user in mind, promoting better health, wellbeing and productivity for occupants.

And to that I would add that it actively contributes towards meeting the UN Sustainable Development Goals.

If we are working within construction and have an opportunity to talk about an organisational, community or even personal journey towards this sort of sustainable built environment, it is our responsibility to do so, using all our professional skill and integrity.

And it *is* almost always a journey: a process of setting targets, measuring progress, making mistakes, learning from those mistakes and trying new approaches.

Admittedly, in many cases, there may be limited appetite in the media for a narrative around the 'journey' towards sustainability. It's too internal. The public does not get excited about basic housekeeping, saving energy or meeting waste targets. And the press is unlikely to rush to cover the launch of a corporate responsibility and sustainability strategy, no matter how noble, unless it's something notable from an already newsworthy brand.

However, in such situations, the emphasis can go more onto Paid and Owned aspects of the PESO model of PR:[33]

- Paid Media: typically promoted or sponsored content in the media and social media, including advertorials; also email marketing to get messages direct to stakeholders.
- Owned Media: your own content, published via a website, blog or in print, film or imagery, where you control the messaging and tell the story in a way you want it told.

Internal communications is also very important, as explained earlier, to boost employee engagement and to support successful recruitment and retention of talent – itself a major problem in many parts of the construction supply chain.

Ultimately, however it is done, what matters most is that we do keep communicating effectively about sustainability, responsibility and sound ethics as simply part of good business leadership in the construction industry in the twenty-first century.

Notes

1. ACS, 'What are the greenhouse gas changes since the Industrial Revolution?': https://www.acs.org/content/acs/en/climatescience/greenhousegases/industrialrevolution.html [Accessed 4 February 2020].
2. UN Secretary-General's remarks on Climate Change, 10 September 2018: https://www.un.org/sg/en/content/sg/statement/2018-09-10/secretary-generals-remarks-climate-change-delivered [Accessed 4 February 2020].
3. Synthesio report 'Social Media Monitoring the Fast-Paced Landscape of Climate Change Conversations', October 2019: https://resources.synthesio.com/social-media-monitoring-climate-change.html [Accessed 4 February 2020].
4. Oxford English Dictionary, definition of climate emergency: https://languages.oup.com/word-of-the-year/2019/ [Accessed 4 February 2020].
5. https://www.climateemergency.uk/blog/list-of-councils/ [Accessed 19 February 2020].
6. Supply Chain Sustainability School: https://www.supplychainschool.co.uk/ [Accessed 4 February 2020].
7. CIPR Influence, 'Five Sustainability Trends For Brands To Watch In 2020', December 2019: https://influenceonline.co.uk/2019/12/06/five-sustainability-trends-for-brands-to-watch-in-2020/ [Accessed 4 February 2020].
8. https://www.ukgbc.org/wp-content/uploads/2019/02/UKGBC-Leading-the-Way-2019.pdf [Accessed 4 February 2020].
9. https://ghgprotocol.org/ [Accessed 4 February 2020].
10. https://ghgprotocol.org/sites/default/files/standards/Corporate-Value-Chain-Accounting-Reporing-Standard_041613_2.pdf [Accessed 4 February 2020].
11. https://www.ukgbc.org/wp-content/uploads/2019/07/Scope-3-Summary-spreads.pdf [Accessed 4 February 2020].
12. https://www.theccc.org.uk/publication/reducing-uk-emissions-2018-progress-report-to-parliament/.

13 BBC, 'Climate change food calculator: what's your diet's carbon footprint?', 9 August 2019: https://www.bbc.co.uk/news/science-environment-46459714 [Accessed 4 February 2020].
14 https://www.gov.uk/government/news/uk-becomes-first-major-economy-to-pass-net-zero-emissions-law [Accessed 4 February 2020].
15 https://www.un.org/sustainabledevelopment/sustainable-development-goals/ [Accessed 4 February 2020].
16 https://www.bioregional.com/one-planet-living [Accessed 4 February 2020].
17 https://www.bioregional.com/resources/build-a-better-future-the-built-environment-and-the-sustainable-development-goals [Accessed 4 February 2020].
18 Barratt Developments plc stakeholder engagement on the UN Sustainable Development Goals: https://www.barrattdevelopments.co.uk/sustainability/building-sustainable-values/stakeholder-engagement/un-sustainable-development-goals [Accessed 4 February 2020].
19 Public understanding of science: lessons from the UK experience, Scidev.net: [Accessed 10 February 2020].
20 Greenwashing, Wikipedia page: https://en.wikipedia.org/wiki/Greenwashing [Accessed 4 February 2020].
21 Sins of Greenwashing, UL: https://www.ul.com/insights/sins-greenwashing [Accessed 4 February 2020].
22 Making a Green Claim, Defra, 21 December 2016: https://www.gov.uk/government/publications/make-a-green-claim/make-an-environmental-claim-for-your-product-service-or-organisation [Accessed 4 February 2020].
23 https://www.asa.org.uk/type/non_broadcast/code_section/11.html [Accessed 4 February 2020].
24 Neither Boastful nor Bashful, UL, 2019: https://collateral-library-production.s3.amazonaws.com/uploads/asset_file/attachment/1563/Neither-Boastful-Nor-Bashful-Innovative_Claims.pdf [Accessed 4 February 2020].
25 Avoid 'Greenblushing': walk and talk at the same time, Dix & Eaton, 2 March 2010: https://www.dix-eaton.com/our-blog/avoid-greenblushing-walk-and-talk-at-the-same-time/ [Accessed 4 February 2020].
26 Science Based Targets initiative – Landsec case study: https://sciencebasedtargets.org/case-studies-2/case-study-land-securities/ [Accessed 4 February 2020].
27 Property Week, Research calls existence of green premium into question, 31 January 2020: https://www.propertyweek.com/news-analysis/research-calls-existence-of-green-premium-into-question/5106219.article [Accessed 4 February 2020].
28 Sell the Sizzle, Futerra: https://www.wearefuterra.com/wp-content/uploads/2018/03/Sellthesizzle.pdf [Accessed 4 February 2020].
29 8 principles for impactful visual communication of climate change, Gavin Lamb PhD, UX Collective, 14 August 2020: https://uxdesign.cc/8-evidence-based-principles-for-impactful-visual-communication-of-climate-change-a266241bc9f8 [Accessed 16 August 2020].
30 Climate Visuals, The evidence behind Climate Visuals: https://climatevisuals.org/evidence-behind-climate-visuals [Accessed 16 August 2020].
31 Chartered Institute of Public Relations, CPD resources: https://cipr.co.uk/content/cpd
32 It's time green marketing grew up, BusinessGreen Editor's blog, 21 August 2012: https://www.businessgreen.com/bg/blog-post/2200154/it-s-time-green-marketing-grew-up [Accessed 4 February 2020].
33 Spin Sucks, 'What is the PESO model?': https://spinsucks.com/what-is-the-peso-model/ [Accessed 4 February 2020].

11 Conclusion

The future of communicating construction

Liz Male and Penny Norton

Where does construction go from here?

A UK construction industry that has been undergoing gradual modernisation now finds itself operating in a rapidly changing and uncertain world, which is being reshaped by technological innovation and disruption, the climate emergency, the COVID-19 pandemic and shifting global geopolitics, with the UK redefining its position following Brexit.

There is the potential for all these factors to influence the future course of the industry. This includes the policy and economic environment driving construction activity, global supply lines for materials and expertise, the processes by which projects are delivered and the technologies and products used, market demands or ultimately, of course, the resulting buildings and structures.

Already, however, clear priorities are evident for UK construction as it plots its way ahead. These all provide significant opportunities for the public relations and communications sector to bring its advice and expertise to support construction organisations as they evolve, adapt, transform and thrive.

Priorities for UK construction communications include:

1. Establishing trust in an age of mistrust

We live in an age of mistrust. Authority is no longer automatic. The actions of all public bodies, businesses and established sources of news and information may be regarded with suspicion, challenged or denigrated.

Construction has also generally done a poor job of fostering trust with consumers and the communities in which it works, and its reputation has been diminished by a string of negative events. Many commentators would say we have the reputation we deserve.

The Grenfell Tower fire in 2017 and Sir Martin Moore-Bick's ensuing Public Inquiry[1] have shown the industry failing on many levels. Grenfell's impact has rippled out, with some 2,000 residential buildings now considered to be covered in some form of dangerous cladding. Their residents are left with properties that are blighted by high refurbishment costs and effectively cannot be sold and mortgaged until they are remedied, as a report from a House of Commons select committee on Housing, Communities and Local Government reported.[2]

The housebuilding sector has attracted negative publicity for issues ranging from poor quality workmanship in new homes[3] to onerous ground rents for leasehold houses.[4] In 2018, top housebuilder Persimmon added further controversy by awarding high bonuses to senior executives, prompting a public and political backlash.[5] The following year, the company responded by introducing a number of remedial measures and processes and initiated an independent review, led by Stephanie Barwise QC of Atkin Chambers, to look at their effectiveness.[6]

Broader measures are being taken to remedy industry failings. Following these concerns, the government announced plans in 2018 to appoint a New Homes Ombudsman to help buyers of new homes gain redress via an alternative dispute resolution process.[7] It is expected that the ombudsman will have powers to award compensation to homebuyers, ban rogue developers and order housebuilders to rectify poor quality work, when the scheme comes into place in 2021.

Dame Judith Hackitt's Independent Review of Building Regulations and Fire Safety,[8] which was initiated following the Grenfell Tower fire, set challenges for UK construction, including one relating to the need to communicate construction product information in a clear and unambiguous way. This challenge was picked up by the Construction Products Association, a trade association, which established a Marketing Integrity Group to establish a scheme to make product information clear, unambiguous, transparent and rigorous. Its call for evidence in 2019 highlighted the strengths and weaknesses of product information and potential areas for improvement[9] (see also the Editor's postscript to Chapter 5).

The TrustMark government-endorsed quality scheme[10] and a suite of new technical and customer service standards for domestic retrofit such as PAS 2030 and PAS 2035 from the BSI[11] are giving greater confidence to customers in the home improvement sector. This is a sector where reputational damage has been done in the past by scammers and rogue traders, poor quality work and a legacy of stop-start government incentive schemes around energy efficiency and renewable energy. TrustMark connects consumers with trusted tradespeople and lays the foundations for high standards of professionalism. PAS 2030 and 2035 provide a specification for the energy efficiency retrofit of existing homes and were introduced following the government-backed Each Home Counts review,[12] which looked at how consumers could be better advised and protected when carrying out this work.

All of these initiatives, and more, are designed to rebuild public trust in the construction industry and its outputs. As PR practitioners in this sector, our job is to seek out and support the best of these, to ensure our behaviours and messaging are aligned.

2. Telling a powerful story

Potential recruits to construction's workforce also have a poor perception of the industry. Surveys and data[13] have highlighted the reluctance of young people to enter an industry where training opportunities are limited, working conditions poor and jobs insecure and highly vulnerable to the vicissitudes of economic boom and bust. This has left construction heavily reliant on an ageing workforce and labour drawn from overseas, prompting the Farmer Review of the UK Construction Labour Model[14] to warn, 'The pure physical capacity of the construction industry to deliver for its clients appears to be in serious long-term decline'.

The industry's future, says the Farmer Review, lies in new business models using manufacturing-led construction methods, alongside reform of the industry's training structure, including the Construction Industry Training Board (CITB).

A reformed or new training body should, it says, be, 'empowered to deliver a more powerful public facing story and image for the holistic "built environment"; process, of which construction forms part. This responsibility should include an outreach programme to schools and draw on existing industry exemplars and the vision for the industry's future state rather than just "business as usual"'.

A 2020 report on the global construction industry by management consultant McKinsey[15] also points to radical change, leading to structures and services being increasingly delivered and marketed as standardised – and branded – products. This customer-centric, branded approach can encompass such features as product and service quality, value, service offerings, timing of delivery and warranties.

Public relations and communications advisors need to gain a full understanding of construction's reform and change in order to convey its new stories to prospective consumers and employees.

3. Valuing the workforce

As explained particularly well in Chapter 4, if the industry is really to change its image it also needs to provide – and be seen to be providing – a more inclusive and supportive environment to drive a more diverse workforce.

Imbalances and prejudice across gender, sexuality, disability, race and colour in construction's workforces are deeply entrenched and proving slow to change. But now activist movements, notably the global Black Lives Matter[16] network, are shining a spotlight on inequality across all aspects of life, in the same way as Swedish activist Greta Thunberg did with environmental issues.

Numerous reports set out ways in which the industry can work to improve diversity, such as The Equality and Human Rights Commission's Equality and diversity: good practice for the construction sector. This publication's action plan highlights communications and media measures, including the promotion of good practice and progress, and the celebration of successful results through rewards for excellence and ambassador programmes.[17]

Significantly, government housing delivery agency Homes England is now showing leadership and driving diversity and inclusion among the businesses it works with. In 2020, it published its first Annual Equality, Diversity and Inclusion Report, setting out its own objectives to promote inclusion, which include requesting diversity and inclusion commitments from organisations on its procurement panels in future.[18]

A more diverse and inclusive workforce also has a part to play in breaking down construction's traditionally macho culture. Mental health is a major health and safety concern for the industry. Between 2011 and 2015 more than 1,400 construction workers died by suicide, and a survey by the Chartered Institute of Building (CIOB) of more than 2,000 construction industry professionals found more than a quarter had thought about taking their own lives in 2019.[19] Job insecurity, long working hours, time away from families, lack of support from human resources and late payment were all found by the survey to be contributing to loss of wellbeing. The report highlighted gender differences even here, with female construction workers often having to work with poor or no toilet facilities and inadequate sanitary conditions and men feeling unable to discuss their mental wellbeing due to 'hyper-masculine' expectations of how they behave.

Action has been long overdue, but such concerns are now receiving the greater attention they merit. Initiatives including the Building Mental Health construction network[20] and charity Mates in Mind[21] both provide useful resources for mental wellbeing, as do the industry's professional bodies.

4. Working with communities

Chapters 2 and 7 explain this in some depth. The ways in which the construction industry engages and works with communities have been evolving, with approaches that empower local people and/or residents growing in popularity. Neighbourhood plans – which are prepared by parish or town councils or a neighbourhood forum – are enabling communities to influence planning in their area, by allowing them to help develop a vision for future buildings and open spaces. A range of community-led housing delivery models put communities in the driving seat in generating homes that meet local needs, particularly in being affordable to local people. One model is via community land trusts, organisations formed by the community to develop and manage homes and other assets, working alone or in partnership with housing associations.

PR professionals working in this area of planning and development communications in England will also need to understand and rapidly adapt to the changes proposed in the latest 'Planning for the Future' white paper.[22]

Processes to increase dialogue and collaboration are increasingly being adopted to improve ways of working with communities. For example, the National Housing Federation, the body representing England's housing associations, last year published its Together with Tenants charter.[23] Developer and landowner Grosvenor has published a community charter, called Positive Space, setting its standard for public engagement, which follows on from its own research identifying a lack of public trust in planning and development.[24]

Developers like Grosvenor and others working in the construction industry are now distinguishing themselves by the ways in which they are working with the communities in which they build. This could become increasingly important

Conclusion   185

in the wake of the COVID-19 pandemic, which has emphasised the need for communities and people to make successful and vibrant places and buildings.

5. Responding to the climate crisis, while acknowledging its complexity

Corporate social responsibility (CSR) has long been the established business term to describe a company's approach to benefitting communities and broader society. But today, it is becoming more common to use the term environmental, social and governance (ESG), which places a definite emphasis on the business response to climate catastrophe, its relationship with its workers and communities and the way in which it operates.

As explained in Chapter 10, a new building or structure can aid the environment through its use of sustainable materials and energy efficiency in use – or it can be harmful in its profligacy with natural resources in construction and use. Buildings and infrastructure can, therefore, present complex and sometimes controversial questions around sustainability.

One example in 2020 is high-rise timber residential construction, which has come under scrutiny because of concerns about potential fire risk, following the Grenfell Tower fire. This issue could impact on the otherwise fast increase in sustainable timber construction, driving the adoption of steel or concrete,[25] which are considered less sustainable.

Sustainability can present similar dilemmas for businesses. At the time of writing, the Architects Declare movement in the UK[26] has attracted almost a thousand signatories, all architectural practices that have signed up to a series of aims around the climate and biodiversity emergency. But some organisations that have signed up have attracted negative publicity for taking on projects that are perceived to conflict with their environmental pledges.[27]

These issues will become no less complex with time, and PR professionals' advice will be essential in ensuring that clients are kept abreast of rapidly evolving concerns arising from areas including policy, research, major campaign groups and local activists.

6. Transforming with technology

Look at an area of construction that is changing, and you are likely to see technology enabling its progress. Chapter 9 really explores this. Building information modelling (BIM) is refining the way construction projects are created, delivered and operated. Modern methods of construction are helping to take processes offsite and into the factory. Drones are changing the way the built environment can be surveyed, and electric and autonomous vehicles and robots are influencing the design of streetscapes and are being used in buildings to access hard-to-reach areas and carry out essential tasks.

All these innovations can help to increase efficiency and improve construction's labour productivity, which traditionally has lagged that of manufacturing and the economy as a whole.[28]

PR as a strategic management discipline

Like the construction industry, the PR world itself is adapting and evolving in response to change. New reputational threats, policy changes and socioeconomic trends demand that communicators in the construction industry need to be seen as operating at a much more informed and strategic level. New media platforms and technologies and very different needs and ways of working with clients are helping to reshape the role of public relations advisors.

Ten top priorities for construction's PR and communications advisors include:

1. Understand the policy context

Central and local government policy have a dramatic impact on construction, driving both direct funding for the provision and upgrading of hospitals, roads and other infrastructure and the framework for construction in planning, Building Regulations and other key areas.

The construction PR team needs to fully understand the policy context and the direction of travel of policy. Construction businesses may not be able to see what is coming over the horizon, but the public relations professional, with their outward-facing role, is well placed to provide advice at a strategic management level on what lies ahead.

2. Understand the purpose and importance of partnerships

In an industry as fragmented as construction, one fundamental skill required by its PR professionals is the ability to work collaboratively and to build networks that achieve positive outcomes for everyone involved. It is not uncommon for PR teams to have to liaise with multiple other organisations in order to get sign-off for a story, for example.

Construction has been urged to become more collaborative for decades. It's the clarion call from seminal industry reports such as the 1994 Latham Report 'Constructing the Team',[29] the 1998 Egan Report 'Rethinking Construction'[30] and almost every industry report published since. Arguably, PR and communications is as good a place to start as any.

But there are also increasing options for more strategic, deeper and long-lasting partnerships. Some have been sparked in response to the COVID-19 crisis – take for example, the new level of collaboration established between industry bodies working together through the Construction Leadership Council[31] to provide guidance and clarity to the construction industry since the lockdown.

Other partnerships are coming from increasing awareness of the need to communicate purpose. According to research findings from media monitoring company alva: 'Smart companies have realised that there is a way for them to demonstrate genuine solidarity, create solutions which unpick tricky social problems and inequalities, and improve their reputation, to boot. That solution is corporate partnerships – in particular, those which involve downing weapons and partnering with companies with whom you're more used to competing'.[32]

3. Be data-proficient

A senior advisory role requires PR professionals to have a good understanding of data strategy and analysis. The ability to scrutinise analytics and build a data-informed picture enables PR professionals to apply evidence-based decision making to inform their strategic advice, helping to minimise waste and inefficiency within the construction industry.

Analytics allow improved evaluation and measurement of public relations activity, enabling advisors to better assess their own effectiveness and report back to clients. The website of the International Association for Measurement and Evaluation of Communication (AMEC) has an essential range of tools and guidance in this area, including a primer on PR planning.[33] These are now non-negotiable areas of expertise for PR practitioners.

4. Keep track of the fast-moving changes in media

Construction is a sector which has traditionally had a very diverse body of specialist publications, with multiple titles covering the multiple trades, professional disciplines and sub-sectors in the supply chain.

As has been mentioned in several of the chapters in this book, that is changing. Currently, we are seeing radical shrinkage and consolidation in the trade press with far less professional journalism, as many outlets become paid-for content distributors. This trend will accelerate post-pandemic – at the time of writing, the COVID-19 crisis has already led to more than 2,000 job cuts across UK news organisations, with more expected to follow.[34]

These changes are exacerbated with the well-documented changes in people's consumption of news via the traditional media – trends that were well underway anyway, even before lockdown. Ofcom analysis of ABC (Audit Bureau of Circulations) figures shows total national newspaper circulation has declined from nearly 22m in 2010 to 9.3m in 2019. A third (35%) of adults claim to consume news through print newspapers. The number using daily freesheets has gone up from 23% last year to 27%, but only 3% read free weekly local newspapers.[35]

TV remains by far the most popular source for news, while around half of adults in the UK get their news not from established news media but from diverse social media sources.[36]

However, Ofcom's analysis suggests there is evidence that UK adults who do use social media for news (45%) are less engaged with the content. Those who use Facebook, Instagram and Twitter are less likely to share/retweet trending news articles, and smaller proportions are clicking on news articles/videos (Facebook/Instagram) or making comments (Twitter/Instagram) compared to 2019.

5. Understand how to build communities of interest

In this pluralist media environment, the public relations professional has to be platform-agnostic, being as comfortable with video and audio content as they are with the written word, and open to social channels as well as local, business, national and even international media outlets.

Paid-for content still has a place in some PR campaigns but requires informed and creative media planning. In the past, paid content may have been simply an 'advertorial' in a publication, but now it is more likely to be a collaboration with key groups, such as Instagram influencers. PR professionals, therefore, need to have their finger on the pulse to identify where communities of interest are located and who they are at any one point in time. If budget is to be spent on paid-for communications, it should be rigorously researched, planned, tracked and evaluated.

Earned content, through traditional media relations, remains important but now demands broader thinking about the right channel for the right time. It also increasingly includes working with bloggers and so called 'nano-influencers'. The outcomes of online earned media campaigns are easily measured via Google Analytics and other digital evaluation tools, and these should be a standard part of today's PR evaluation toolkit.

Shared and social content also continues to be extremely important and requires public relations expertise to understand and monitor networks and to tailor content.

Owned content is the area that is seeing the greatest growth, as companies become their own publishers and extend their reach using SEO-savvy content marketing. This offers opportunities for PR practitioners to confirm their expertise in the content marketing space – for example, we should be celebrating the direct and measurable impact our work has on the domain authority and search rankings of our clients' websites.

With the growth in digital platforms, audiences can also be more finely targeted. Working with small influencer groups can often achieve marketing objectives faster and more effectively than a major advertising campaign and at a fraction of the cost. Geo-targeting practices allow content and communications to have a tight geographical focus. This is beneficial not only when working within the nations of the United Kingdom but in reaching geographical areas targeted for key construction projects or city regions.

The staple output of the PR company used to be the single-side-of-A4-paper press release, accompanied by a photo. Now, an advisor could be putting across a campaign message in a TikTok video, podcast, Alexa skill or white paper. The range of media and formats now in the communications armoury requires advisors to be as comfortable working with video and voice as they are with words. These trends are only expected to accelerate.

6. Understand the impact of AI

Technological innovation is driving further change. Content generated via artificial intelligence (AI) is rapidly improving in quality, and the range of data-driven, machine learning and AI-enabled tools for PR practitioners increases month by month, particularly in areas such as media monitoring and evaluation.

The CIPR Artificial Intelligence (#AIinPR) panel[37] was founded in February 2018 to explore the impact of artificial intelligence on public relations and the wider business community. The panel has established an international reputation as a centre of excellence for promoting knowledge and understanding of AI.

In the CIPR guide on AI and the media,[38] author Andrew Bruce Smith sets out the extent to which AI and automation has already impacted the media industry and the implications of this for PR practitioners. One example looks at the rise of machine-generated content:

> AI and machine learning may play an increasing role in generating, testing and identifying the best PR content. In July 2019, JP Morgan Chase signed a five-year contract with Persado,[39] a software start-up that uses artificial intelligence to write marketing copy, following a successful pilot. In doesn't take much of a stretch to see that similar technology could be applied in the realm of public relations content.

In just one example linked to housebuilding, heavily personalised content, backed up with technology that instantly sees and learns what works and what doesn't, has huge scope to transform the way new homes are marketed and sold, just as this technology is already doing with other B2C markets.

The PR professional in construction should be informed about these new developments and alert to the complex ethical challenges that they also create.

7. Get smarter at demonstrating return on investment

For decades, the argument has raged about whether PR can be responsible for generating sales or its value judged by its impact on direct commercial targets and the bottom line. In 'traditional' PR terms, this is often very difficult to measure.

But for PR practitioners using the full range of techniques available in the PESO model, it is perfectly possible to combine creative excellence, increased awareness and enhanced reputation with other results closely linked to business outcomes.

Perhaps more importantly than ever before, as the UK economy enters a period of recession expected to be one of the harshest in history, the construction industry deserves PR that moves beyond 'vanity metrics' into campaigns that drive sales qualified leads and sustained company growth. Expect to see massively increased demand for improved evaluation and measurement of return on investment. Chapter 6 explains how this has always been part of the picture for PR practitioners in the building products sector. We can expect to see more sophisticated approaches to this extend into many other parts of the supply chain too.

8. Increase the power of visual story-telling

For what is a largely visually-oriented industry, some parts of construction were surprisingly slow to see the potential for communicating their value through visual story-telling.

A range of stand-out initiatives have helped to address this, including the decision to bring in artists-in-residence to major projects. This has led to the creation of the hugely atmospheric paintings created by Robert Mason of the construction of Broadgate in London in 1989/90,[40] Julie Leonard's work for Crossrail in 2014–15[41] and Patricia Cain's award-winning work over three years recording the construction of Zaha Hadid-designed Riverside Museum in Glasgow.[42]

Other examples are the highly successful Art of Building competition[43] run every year by the CIOB as an international showcase for the very best digital photography of the built environment, as well as the fast-growing YouTube phenomenon of The B1M.[44] There is still so much more potential for construction to explore in this area.

As PR professionals in an industry with a poor track-record in diversity, equality and inclusion, we should also be aware of the importance of using local, ethnic or gender-balanced voices in our communications, including photography, video and multi-media story telling. We should be empowering those voices who are best able to tell the stories we need to communicate.

9. Promote leadership

Many forms of current content are tagged with the term 'thought leadership', but in practice little of it lives up to that description. The term has become devalued, with most content in this space providing plenty of thought but relatively little leadership.

True thought leadership entails building a community of interest, moving that forward and developing it into a movement of people and organisations who wish to follow. It is not a quick and easy undertaking, and construction has traditionally lacked the role models and high profile leadership that are commonly observed across other major industries.

Both small and large companies within construction could be taking a leadership stance, with the help of PR professionals to prepare the ground. That groundwork includes identifying opportunities to make a difference, helping a leader establish their position at the forefront of a movement, engaging with others on collaboration and communicating every step along the way.

10. Establish and maintain expert status

All of these factors emphasise PR's role as a strategic management function. The PR professional should be operating at board level as comfortably as at the technical, tactical level, being the construction organisation's head of corporate reputation, highlighting opportunity and advantage and safeguarding against risk.

Today's core professional skills are no longer based around writing a press release. We should not forget the research and plan phase of what we do. And we must never forget to measure our work.

We must have an understanding of current affairs, dexterity in data analysis and the ability to create communities and keep those communities engaged. We need creativity to communicate the amazing and admirable work of so many parts of the construction sector, and we must measure the impact we make in a way that demonstrates genuine return on investment.

Every PR professional in this sector should have, or be working to achieve and maintain, such expert status. Ongoing continuous professional development (CPD) is also vital, as is the move towards gaining chartered status via the CIPR to demonstrate total mastery of PR ethics, strategy and leadership. Any student reading this book and any public relations professional already in the industry with these attributes on their CV will have highly marketable skills into the future.

Notes

1 Grenfell Tower Inquiry https://www.grenfelltowerinquiry.org.uk/ [Accessed 8 August 2020].
2 House of Commons Select Committee on Housing, Communities and Local Government, Report, Cladding: progress of remediation https://publications.parliament.uk/pa/cm5801/cmselect/cmcomloc/172/17202.htm [Accessed 24 July 2020].
3 Why are Britain's new homes built so badly? *The Guardian*, 11 March 2017 https://www.theguardian.com/money/2017/mar/11/why-are-britains-new-homes-built-so-badly [Accessed 24 July 2020].
4 Leasehold houses and the ground rent scandal: all you need to know. *The Guardian*, 25 July 2017 https://www.theguardian.com/money/2017/jul/25/leasehold-houses-and-the-ground-rent-scandal-all-you-need-to-know [Accessed 30 July 2020].
5 Persimmon boss asked to leave amid outrage over bonus, *The Guardian*, 7 November 2018 https://www.theguardian.com/business/2018/nov/07/persimmon-boss-asked-to-leave-amid-ongoing-outrage-over-bonus [Accessed 24 July 2020].
6 Persimmon Plc, Independent Review https://www.persimmonhomes.com/corporate/corporate-responsibility/independent-review [Accessed 28 July 2020].
7 Ministry of Housing, Communities and Local Government, technical consultation: Redress for purchasers of new build homes and the New Homes Ombudsman, June 2019 https://www.gov.uk/government/consultations/redress-for-purchasers-of-new-build-homes-and-the-new-homes-ombudsman [Accessed 24 July 2020].
8 Independent Review of Building Regulations and Fire Safety: Final Report – Building a Safer Future, May 2018 https://assets.publishing.service.gov.uk/government/uploads/system/uploads/attachment_data/file/707785/Building_a_Safer_Future_-_web.pdf.
9 Construction Product Information Survey – Our initial findings, Construction Products Association in collaboration with NBS https://www.constructionproducts.org.uk/publications/technical-and-regulatory/construction-product-information-survey/ [Accessed 24 July 2020].
10 TrustMark https://www.trustmark.org.uk/ [Accessed 8 August 2020].
11 British Standards Institution (BSI) PAS standards for retrofit: https://www.bsigroup.com/en-GB/about-bsi/media-centre/press-releases/2019/august/bsi-establishes-new-framework-for-the-installation-of-energy-efficiency-measures-in-existing-dwellings/ [Accessed 8 August 2020].

12 Each Home Counts review http://www.eachhomecounts.com/ [Accessed 8 August 2020].
13 Construction drops out of top 10 jobs for young people, Building, 27 September 2018 https://www.building.co.uk/news/construction-drops-out-of-top-10-jobs-for-young-people/5095790.article [Accessed 30 July 2020].
14 Construction Leadership Council. Farmer, M. The Farmer Review of the UK construction labour model, Modernise or die: time to decide the industry's future, October 2016 https://www.constructionleadershipcouncil.co.uk/wp-content/uploads/2016/10/Farmer-Review.pdf [Accessed 24 July 2020].
15 The next normal for construction: How disruption is reshaping the world's largest ecosystem. McKinsey & Company, June 2020 https://www.mckinsey.com/~/media/McKinsey/Industries/Capital%20Projects%20and%20Infrastructure/Our%20Insights/The%20next%20normal%20in%20construction/executive-summary_the-next-normal-in-construction.pdf [Accessed 30 July 2020].
16 Black Lives Matter https://blacklivesmatter.com/ [Accessed 8 August 2020].
17 Equality and diversity: good practice for the construction sector, A report commissioned by the Equality and Human Rights Commission, May 2011 https://www.equalityhumanrights.com/sites/default/files/ed_report_construction_sector.pdf.
18 Annual Equality, Diversity and Inclusion Report 2020/21, Homes England, 29 July 2020. https://assets.publishing.service.gov.uk/government/uploads/system/uploads/attachment_data/file/904576/ED_I_Report.pdf [Accessed 30 July 2020].
19 Chartered Institute of Building. Understanding mental health in the built environment, May 2020 https://policy.ciob.org/wp-content/uploads/2020/05/Understanding-Mental-Health-in-the-Built-Environment-May-2020-1.pdf [Accessed 24 July 2020].
20 Building Mental Health https://www.buildingmentalhealth.net/ [Accessed 8 August 2020].
21 Mates in Mind https://www.matesinmind.org/ [Accessed 8 August 2020].
22 Planning for the Future, MHCLG, August 2020 https://assets.publishing.service.gov.uk/government/uploads/system/uploads/attachment_data/file/907647/MHCLG-Planning-Consultation.pdf [Accessed 8 August 2020].
23 Together with Tenants, National Housing Federation, July 2019 https://www.housing.org.uk/our-work/together-with-tenants/ [Accessed 27 July 2020].
24 Positive space our community charter, Grosvenor Britain and Ireland, June 2020 https://www.grosvenor.com/Grosvenor/files/88/888c96aa-7f04-4cbd-9745-7fd47fd8128c.pdf [Accessed 30 July 2020].
25 Grenfell fears prevent timber building boom, BBC News, 25 May 2020 https://www.bbc.co.uk/news/business-52771270 [Accessed 29 July 2020].
26 Architects Declare https://www.architectsdeclare.com/ [Accessed 8 August 2020].
27 Foster+Partners' Architects accused of climate betrayal over new airport in Saudi Arabia. *The Times* https://www.thetimes.co.uk/article/foster-and-partners-architects-accused-of-climate-betrayal-over-new-airport-in-saudi-arabia-bf58pz2v5 [Accessed 27 July 2020].
28 Construction statistics, Great Britain: 2018 ONS https://www.ons.gov.uk/businessindustryandtrade/constructionindustry/articles/constructionstatistics/2018 [Accessed 28 July 2020].
29 Designing Buildings Wiki on Latham Report https://www.designingbuildings.co.uk/wiki/Latham_Report.
30 Designing Buildings Wiki on Egan Report https://www.designingbuildings.co.uk/wiki/Egan_Report_Rethinking_Construction
31 Construction Leadership Council http://www.constructionleadershipcouncil.co.uk/ [Accessed 16 August 2020].

Conclusion 193

32. Corporates and the (newfound) spirit of partnership, Influenceonline.co.uk, 13 August 2020 https://www.influenceonline.co.uk/2020/08/13/corporates-and-the-newfound-spirit-of-partnership [Accessed 13 August 2020].
33. Primer: Introduction to PR planning, Moroney, G, head of insight and strategy ENGINE Mischief and partner ENGINE, on behalf of AMEC, November 2019 https://amecorg.com/wp-content/uploads/2019/11/AMEC-PRIMER-planning-guide-MM2019.pdf [Accessed 29 July 2020].
34. Press Gazette, Covid-19 crisis leads to more than 2,000 job cuts across UK news organisations, 14 August 2020 https://www.pressgazette.co.uk/covid-19-crisis-leads-to-more-than-2000-job-cuts-across-uk-news-organisations/ [Accessed 14 August 2020].
35. Press Gazette, Ofcom survey News consumption via social media and Google is down, 13 August 2020 https://www.pressgazette.co.uk/ofcom-uk-news-consumption-survey-news-consumption-via-social-media-and-google-is-down/ [Accessed 13 August 2020].
36. News consumption in the UK, Ofcom, 24 July 2019 https://www.ofcom.org.uk/research-and-data/tv-radio-and-on-demand/news-media/news-consumption [Accessed 29 July 2020].
37. Chartered Institute of Public Relations (CIPR) Artificial Intelligence Panel https://cipr.co.uk/CIPR/Our_work/Policy/CIPR_Artificial_Intelligence_in_PR_panel.aspx [Accessed 8 August 2020].
38. CIPR's AI in PR Primer, 'The impact of AI in Media and PR', Andrew Bruce Smith https://cipr.co.uk/CIPR/Learn_and_develop/Resources/CIPR/Learn_Develop/Resources/CIPR_Resources.aspx?hkey=9d6f264b-9f1e-4234-b4ce-7356fab9fc2d [Accessed 8 August 2020].
39. Persado website https://www.persado.com/ [Accessed 8 August 2020].
40. Robert Mason *Broadgate Paintings and Drawings* 1989–90, Yale University Press; 1st edition (1 May 1990).
41. Construction artwork from artist in residence, Crossrail https://www.crossrail.co.uk/route/art-on-crossrail/cosntruction-artwork-from-artist-in-residence [Accessed 16 August 2020].
42. Patricia Cain, Drawing on the Riverside http://www.patriciacain.com/drawing-(on)-riverside.html [Accessed 16 August 2020].
43. The Art of Building https://www.artofbuilding.org/ [Accessed 16 August 2020].
44. The B1M video channel for construction https://www.theb1m.com/ [Accessed 16 August 2020].

Glossary

3D printing The computer-controlled sequential layering of building materials (such as concrete or composite materials) to create three-dimensional shapes. It is particularly useful for prototyping and for the manufacture of geometrically complex components.

ABC certificate Certification from the Audit Bureau of Circulation.

Above the line Promotional methods that cannot be directly controlled by the company selling the goods or service, such as television or press advertising. See also *Below the line*.

Accessibility Freedom for people to take part, including elderly and disabled people, those with young children and those who may encounter discrimination.

Accreditation Formal quality assurance approval via a scheme accredited by the UK Accreditation Service (UKAS).

AEC The architecture, engineering and construction industry.

Affordable housing Social rented, affordable rented and intermediate housing, provided to eligible households whose needs are not met by the market. Eligibility is based on local incomes and local house prices (UK).

Agile working A methodology originally developed within the IT sector but now widely used to improve project management and to develop faster and more responsive working practices.

Allocated site A site with potential for development and allocated as such in a local authority's Local Plan.

AMEC International association for the measurement and evaluation of communication.

Amenity A positive element that contributes to the overall character or enjoyment of an area. For example, open land, trees, historic buildings or less tangible factors such as tranquility. Residential amenity considerations may include privacy (overlooking), overbearing impact, overshadowing or loss of daylight/sunlight.

Analytics The data collected about visitors to a website, used to understand user behaviour. Also used as shorthand for Google Analytics, a specific service provided by Google.

App (application) A type of software programme that can be downloaded onto a computer, tablet or smartphone.

Appeals The process whereby a planning applicant can challenge a decision, usually refusal of planning consent. Appeals can also be made against the failure of a planning authority to issue a decision within a given time and against conditions attached to a planning permission. In England and Wales, appeals are processed by the Planning Inspectorate.

Approved document Approved documents provide guidance for how the Building Regulations can be satisfied in common building situations.

Artificial intelligence The use of computer systems to perform tasks normally requiring human intelligence, such as visual perception, speech recognition, decision-making and translation between languages.

Audience/target audience A specified group defined for marketing purposes.

Augmented reality A way of overlaying additional computer-generated visuals or other information to objects and physical assets in the real world. Used to enhance natural environments or situations and offer perceptually enriched experiences.

Average session time The total duration of all sessions (in seconds) spent on a website, divided by the total number of sessions.

B2B Business-to-business PR.

B2C Business-to-consumer PR.

Barcelona Principles A series of statements to guide best practice in PR measurement that were endorsed after a vote of global delegates at the AMEC European Measurement Summit in 2010.

Below the line Promotional tactics that can be controlled by the company selling the goods or service, such as in-store offers and direct selling. See also *Above the line*.

BIM See *Building Information Modelling*.

Blockchain A system in which a record of transactions made in bitcoin or another cryptocurrency are maintained across several computers that are linked in a peer-to-peer network.

Blog An abbreviation of weblog: a journal that is available online and is updated by the owner regularly.

BME (Black and Minority Ethnic) Terminology used to describe people of non-white descent. BAME (Black, Asian and Minority Ethnic) is also often used.

Bot A software application that runs automated tasks over the Internet.

Bounce rate The percentage of visitors to a website who navigate away from the site after viewing only one page.

Brand The tangible and intangible attributes of a product or organisation that create an image in the public mind.

Brand ambassador A person who is either paid or voluntarily chooses to represent a brand in a positive light and to help amplify its communications.

Brand associations The knowledge and feelings that consumers associate with a brand name.

Brand collaboration   A strategic tool used to gain higher profits through an alliance with another powerful brand name.
Brand essence   The intangible characteristic which defines a brand.
Brand tone of voice   A description of how the brand speaks to consumers.
Brexit   The process of the UK's exit from the European Union.
Broadband   A high-speed internet connection.
Brownfield land   Land which is or was occupied by a permanent structure.
Browser   Computer software which can be used to search for and view information on the internet.
Builders' merchant   A retailer (trade counter) which sells building products to the construction industry.
Building Control   The process of checking compliance with the Building Regulations. Building control services can be delivered by a local authority building control department or via a private sector Approved Inspector.
Building Information Modelling (BIM)   An intelligent 3D model-based process used to plan, design, construct and manage buildings and infrastructure.
Building Regulations   Most building work carried out in England must comply with the Building Regulations 2010. Similar regulations apply in Scotland, Wales and Northern Ireland. Building Regulations approval is different to planning permission and listed building consent. See also *Building standards*.
Building standards   A set of standards established and enforced by local government for the structural safety and broader performance of buildings. See also *Building Regulations*.
Buying off-plan   Buying a new home before it is built.
CAD   Computer-aided design.
Call-in   In the UK, the Secretary of State for Housing, Communities and Local Government can order that a planning application or Local Plan is taken out of the hands of a local authority. The application will then be subject to a public inquiry presided over by a Planning Inspector who will make a recommendation to the Secretary of State.
Carbon footprint   A way of presenting the total greenhouse gas emissions caused by an individual, event, organisation, service or product.
Case study   A detailed explanation of a real-life experience which tells a story that illustrates key principles, features, benefits and lessons learned (e.g. use of a building product).
CGI (computer-generated imagery)   Special visual effects created using computer software.
CIPR (Chartered Institute of Public Relations)   The professional body for the UK public relations industry, providing training and events, news and research.
Circular economy   An approach designed to eliminate waste and to encourage reuse, repair, refurbishment and recycling. A circular economy is an alternative to a traditional linear economy (make, use, dispose) in which we keep resources in use for as long as possible, extract the maximum value from them

198 *Glossary*

whilst in use, then recover and regenerate products and materials at the end of each service life.

Circulation The total number of copies of newspapers, magazines or other print publications distributed by a specific print publication.

Client In construction terms, the client is usually the person or organisation who has commissioned, funded and/or will own the building or construction project.

CNC machine Computer numerical control machine, often used for cutting and carving construction materials, taking instructions from CAD software.

Colour separations The practice of (mostly) trade publications asking for payment before publishing a press release or other editorial coverage. In effect, a form of advertising, but often without the research, media planning or evaluation that accompanies most organisations' display advertising campaigns.

Commercial property Properties which include retail, office buildings, hotels and service establishments although often used specifically in relation to office property.

Community benefits Aspects of a proposed development which bring about social, economic or environmental benefits. Community benefits may be put in place to mitigate the impact of development.

Community engagement Activities undertaken to establish effective relationships with individuals or groups within a defined community.

Community involvement Effective interactions between applicants, local authorities, decision-makers and individual and representative stakeholders to identify issues and exchange views on a continuous basis.

Community relations Social outreach programmes designed to build relations and foster understanding of the role of the business to neighbours in the local community.

Completion Practical Completion is a contractual term used in the building contract to signify the date on which a project is handed over to the client. The date triggers a number of contractual mechanisms.

Consultation The process of sharing information and promoting dialogue between local planning authorities, applicants, individuals or civic groups, with the objective of gathering views and opinions on planning policies or development proposals.

Consultation fatigue The reluctance to take part in consultation, usually as a result of excessive past consultation or lack of demonstrable results from previous consultation.

Contech Construction technology.

Continuing Professional Development (CPD) Continuing Professional Development refers to the process of tracking and documenting the skills, knowledge and experience gained both formally and informally in a job, beyond any initial training or qualifications.

Contractor An organisation appointed by a client to undertake the construction work. May often then sub-let parts of the work to specialist subcontractors or tradespeople.

Glossary 199

Corporate Social Responsibility (CSR)   The recognition that a company or organisation should take into account the effect of its social, ethical and environmental activities on its staff and the community around it. See also *ESG*.

Coronavirus   See *COVID-19*.

COVID-19   Global pandemic which took hold in 2020, resulting in significant economic and social impacts.

Crisis communications   A damage limitation communications process used by organisations use when experiencing a crisis.

Design for Manufacture and Assembly (DfMA)   See *Modern Methods of Construction (MMC)*.

Developer   A developer constructs, redevelops or refurbishes buildings, usually in order to make a profit. Often, a developer is the client, in construction terms.

Devolved authorities   UK regional or country governments with their own powers to operate independent of Westminster and Whitehall on agreed matters.

Digital twin   A virtual model or replica of assets, processes, systems and other entities. While BIM may deliver a digital model of the complete physical asset, the digital twin concept envisages a constant bi-directional digital connection between the digital model and its physical manifestation.

Direct marketing   Advertising and printed promotional material such as brochures, flyers and mailshots sent directly to customers.

Discussion board (online)   An online 'bulletin board' where individuals can post messages and respond to others' messages.

Discussion forum (online)   See *Discussion board (online)*.

Discussion group (online)   See *Discussion board (online)*.

District plan   A document outlining a local authority's plans for the management of land.

Dwell time   The amount of time that a user spends on a website.

Earned media   In the PESO model, earned media is the equivalent of traditional media relations: getting an organisation's news and information printed in online and hardcopy newspapers, trade publications or in the broadcast media.

Elevation   In architectural terms, an elevation is a two-dimensional drawing of a building's façade.

Embargo   The sharing of unannounced, relevant information between a communications professional and the media, with agreement that the information cannot be published before an agreed upon time and date.

Emissions   Typically used as shorthand for carbon emissions or other greenhouse gases.

Energy efficiency   Energy efficiency relates to the thermal performance of a building and how much energy is used to run it. Increasing energy efficiency not only allows individuals and organisations to reduce their capital and operational costs, it can also help lower fuel consumption and so reduce the emission of greenhouse gases and help prevent climate change.

Energy Performance Certificate (EPC) Energy efficiency ratings for residential properties (UK).

Environmental, Social & Governance (ESG) Environmental, social and governance (ESG) criteria are a set of standards for a company's operations, usually used within the context of socially-responsible investments.

Equality, Diversity & Inclusion (EDI) Equality in the workplace means equal job opportunities and fairness for employees and job applicants. Diversity is the range of people in the workforce. For example, this might mean people with different ages, religions, ethnicities, people with disabilities and both men and women. It also means valuing those differences. An inclusive workplace means everyone feels valued at work.

Evaluation The continuous process of measuring the impact of a PR campaign.

Exclusive A news story offered by a PR practitioner to a single newspaper title, radio, website or TV station.

Fake news A recent term used to describe false stories that appear to be news and spread on the internet or using other media.

First fix/second fix First fix is a short-hand term used to describe the processes that are undertaken during construction works up to the point of applying internal surfaces – typically plaster. It is normally used in relation to the work of specific trades such as carpenters, plumbers and electricians. Second fix takes place after the internal surfaces have been applied – for example, fitting of internal doors, skirting, architraves, handrails, fixtures and fittings.

Fit-out A term used to describe the process of making interior spaces suitable for occupation. It is often used in relation to office developments, where the base construction is completed by the developer and the final fit out by the occupant.

Followers People who choose to follow a social media account.

Forward features A list of editorial topics created by publications to plan content coming up throughout the year. An editorial calendar of forward features usually includes descriptions, deadlines for submission of content and publication dates.

Foundations Foundations provide support for structures, transferring their load to layers of soil or rock that have sufficient bearing capacity and suitable settlement characteristics to support them.

Frameworks Clients that are continuously commissioning construction work use frameworks to invite tenders from suppliers of goods and services to be carried out over a period of time on a call-off basis, as and when required.

GDPR (The General Data Protection Regulation) An EU law on data protection and privacy.

Geospatial data Data and information about objects, infrastructure, events or phenomena that have a location on the surface of the earth. Data is held within Geographic Information Systems (GIS) and GIS databases provide

Glossary 201

geolocated access to names, addresses, uses and information about roads, bridges, buildings and other urban features.

Geo-targeting   The process of targeting a marketing or advertising campaign at a limited set of consumers based on their physical location.

Golden thread   A key recommendation within the Hackitt Review, the 'golden thread' seeks to extend fire safety best practice beyond fire professionals to wider society, relaying key information about a building through a chain of stakeholders, from architects to fire engineers through to building safety managers to end-users. The ultimate aim is to arm all building users with information to protect their health and safety.

Google Analytics   A web analytics service which tracks and reports website traffic.

Grassroots engagement   Bottom-up, rather than top-down decision making, sometimes considered more natural and spontaneous than more traditional power structures.

Greenbelt   A designation for land around certain cities and large built-up areas which aims to keep the land permanently open or undeveloped.

Greenfield   Previously undeveloped land.

Greenhouse gas   A greenhouse gas is a gas that absorbs and emits radiant energy within the thermal infrared range. Greenhouse gases cause the greenhouse effect on planets. The primary greenhouse gases in Earth's atmosphere are water vapor, carbon dioxide (CO_2), methane, nitrous oxide and ozone.

Grenfell   Grenfell Tower was a high-rise residential building in North Kensington, London which was destroyed by fire in June 2017, causing 72 deaths. It was one of the UK's worst modern disasters. A Public Inquiry and other investigations (such as the Hackitt Review) are likely to have significant impacts on the ways in which certain buildings are procured, designed, built and maintained in future.

Groundworks   The process of preparing a construction site for foundations.

Hackitt Review   The Independent Review of Building Regulations and Fire Safety was announced by the UK government in July 2017 following the Grenfell Tower tragedy and was led by Dame Judith Hackitt. It examined building and fire safety regulations and related compliance and enforcement with the focus on multi-occupancy high-rise residential buildings. An interim report was published on 18 December 2017, and the final report was published on 17 May 2018.

Hard hats   Safety helmets worn by people working on construction sites.

Hard to reach groups   Those groups of society which it is particularly difficult to communicate with through the usual means.

Hashtag   A hashtag – written with a # symbol – is used to index keywords or topics in social media.

High-viz   High visibility, brightly coloured workwear used to help protect people working on or visiting a construction site.

Hoardings   Temporary fences around a building or structure under construction or repair.

HTML (Hypertext Markup Language) A standardised language of computer code, imbedded in 'source' documents behind all web documents, containing textual content, images, links to other documents and formatting instructions for display on the screen.

Housing association (HA) A common term for independent, not-for-profit organisations which work with local authorities to offer homes to specific demographics at a reduced cost.

Impressions In online analytics, a term to describe the number of times that content is displayed (for example on social media or a website).

Infrastructure Infrastructure projects – typically energy, transport, utilities and digital communication projects requiring civil engineering expertise – are the backbone of a successful modern economy. Well-designed infrastructure projects have long-term economic benefits, contributing to economic growth and productivity.

Influencers Bloggers, journalists and companies who are thought-leaders in their industry and recognised by customers as people to trust.

Infographic An image that breaks down the facts or messages around a key subject into simple graphics.

In-house Staff within a company or organisation (in PR, the term is used to differentiate from consultancies).

Inspector's report A document produced by an independent inspector from the Planning Inspectorate. It assesses the soundness and robustness of planning documents (UK).

Integrated campaign Use of multiple marketing communications channels such as online, print, TV and radio, B2B, direct marketing, video and advertising.

Integrated communications The linking together of all forms of communications and messages.

Issue A matter of concern with potential to become a crisis.

Issues log A simple list or spreadsheet that helps managers track the issues that arise in a project and prioritise responses to them.

Issues management Ongoing activity that includes studying public policy matters and other societal issues of concern to an organisation.

Key performance indicator (KPI) A set of values against which to measure success; must be defined to reflect objectives and strategy and be sufficiently robust for the measurement to be repeatable. KPIs can be presented as a number, ratio or percentage.

Landing page The first page that a web user 'lands' on after they click a link or online advertisement.

Lean construction An adaptation of lean production techniques (first developed by Toyota) applied to the construction industry. Very broadly it can be characterised as techniques aimed at maximising value and minimising waste.

Likes An expression of approval to something posted on social media.

Listed building Buildings of special architectural or historic interest. A listed building may carry certain obligations and restrictions governing its use, repair and maintenance.

Lobbying The process whereby individuals, civic groups or commercial organisations seek to influence planning decision-makers by employing a variety of tactics.

Local Authority (LA) Local government in England consists of five different types of local authorities: single-tier (metropolitan authorities, London boroughs, unitary or shire authorities) and two-tier authorities (county council and borough/district council). The nearly 400 local authorities are responsible for a range of services for people and businesses in a defined area and are made up of permanent council staff, council officers and elected councillors. See also *Local planning authority*.

Local Plan The main planning policy document for a local authority area. A Local Plan's 'development plan' status means that it is the primary consideration in deciding planning applications (UK).

Local planning authority The public authority whose duty it is to carry out specific planning functions for a particular area. In the UK, this includes borough/district councils, London borough councils, unitary authorities, county councils, the Broads Authority, the National Park Authority and the Greater London Authority.

Localism The 2011 UK Localism Act devolved greater powers to local government and neighbourhoods and gave local communities additional rights over planning decisions.

Low carbon homes Houses/apartments specifically engineered to achieve greenhouse gas reduction.

Main contractor The lead contractor on a project, often referred to as a major contractor or Tier 1 contractor. On major projects, there may be more than one major contractor appointed to a project, each with its own supply chain.

Marketing channels The people, organisations and activities necessary to transfer the ownership of goods from the point of production to the point of consumption.

Masterplan A document outlining the overall approach to the layout of a development.

Media relations Communications with the news media on behalf of an organisation.

Messages Agreed words or statements that an organisation intends to communicate to its audiences.

Messaging strategy A set of foundational points that are aligned with a company's goals and overall brand messaging.

Metrics The use of data to gauge the impact of activity, for example on a website or on social media, used to gather information about how a brand, product or company-related topic is perceived.

MHCLG See *Ministry of Housing, Communities and Local Government (MHCLG)*.

Microsite   A small auxiliary website designed to function as a supplement to a primary website.

Micro-targeting   Direct marketing datamining techniques that involve predictive market segmentation.

Ministry of Housing, Communities and Local Government (MHCLG)   The UK Government department with responsibilities for housing, planning and development.

Mixed use   Development projects that comprise a mixture of land uses.

Mixed use development   Developments constituting more than one use type.

Modern Methods of Construction (MMC)   MMC centres around the use of offsite construction techniques that can benefit from factory conditions and mass production techniques. MMC can include factory-manufactured panellised units which are then assembled on-site, modular and volumetric construction 'pods' that are completed in controlled factory conditions prior to transport to site, hybrid techniques and component-level MMC such as floor or roof cassettes, pre-cast concrete foundation assemblies, pre-formed wiring looms, mechanical engineering composites and other innovative techniques to speed up the process of construction.

Modern slavery   Modern slavery is the recruitment and harbouring of humans through the use of force, abuse, deception and coercion for exploitative purposes. The construction sector is particularly vulnerable to modern slavery because of its extended supply chain, opaque procurement processes, skills shortage and high demand for migrant labour. This means that supply chains can conceal human rights abuses and exploitation.

Modular construction   See *Modern Methods of Construction (MMC)*.

Monitoring   Regular measurement of progress towards targets, aims and objectives. Also involves scrutiny, evaluation and, where necessary, changes in policies, plans and strategies.

Multidisciplinary   In construction terms, a consultancy that provides a wide variety of professional services such as engineering, architecture, planning, building services etc.

National Planning Policy Framework   A comprehensive document which sets out the Government's national planning requirements, policies and objectives. It replaces Planning Policy Statements, Planning Policy Guidance and Circulars. The NPPF is a material consideration in the preparation of LDDs and when considering planning applications.

Nationally Significant Infrastructure Project (NSIP)   A project of a type and scale defined under the Planning Act 2008 in relation to major infrastructure developments – usually energy, transport, water and waste. These projects require a single development consent which follows a different procedure to that of standard planning applications.

Neighbourhood forum   A community group that is designated to take forward neighbourhood planning in areas without parishes.

Net zero   This refers to achieving an overall balance between the carbon emissions produced in the construction and operation of all built assets and the carbon emissions taken out of the atmosphere through natural or engineered methods.

News angle/news hook Information which is new, important, different or unusual about a specific event, situation or person and which makes it newsworthy.

NIMBY (Not In My Back Yard) Used in relation to those who oppose development in the vicinity of their homes for purely selfish reasons.

Objection A written representation made to a local planning authority by an individual, civic group or statutory consultee in response to local plan proposals or a planning application.

Offsite construction See *Modern Methods of Construction (MMC)*.

Ombudsmen Independent professional bodies that investigate complaints on behalf of customers.

Online consultation Consultation which takes place via website, email, social media or other online means.

Online forum See *Discussion forum (online)*.

OpEd A newspaper article written by an expert and positioned on the page opposite the editorial page.

Open house (or open viewing) A process, normally managed by an estate agent, whereby several potential purchasers are given a set time during which they may view a property for sale instead of separate, private viewings.

Open/public meeting A meeting (for example to launch a consultation or present and discuss a development proposal) which is open to all.

Organic search The method of finding a website by entering search items into a search engine.

Outline application An application for planning permission which does not include full details of the proposal. Outline consent approves the principle of development and detailed consent is provided at a later stage.

Outputs, outtakes, outcomes Three ways to measure a PR campaign. Outputs are the most basic form of measurement – what was produced and how well it was produced. Outtakes focus on who was reached and the awareness created. Outcomes measure the results of the PR activity, including changes to behaviour and shifts in attitude.

Owned media In the PESO model, owned media refers to an organisation's own content marketing. Typically, it includes an organisation's website content, online articles, blog posts, infographics, videos and other materials designed to connect with customers.

Page views The number of times a web page was viewed.

Paid media In the PESO model, paid media refers to traditional advertising, online and social media advertising, sponsored content and email marketing created as part of a PR campaign.

Parish/town council An elected local government body which provides a limited range of local public services and makes representations on behalf of the community to other organisations.

Passivhaus An international design standard for ultra low energy buildings.

Perception audit/survey A strategic marketing research technique to provide a baseline indication of existing attitudes against which to measure future marketing efforts.

PESO The PESO model merges four types of media: Paid, Earned, Shared and Owned.

PEST analysis (or PESTEL) A framework of macro-environmental forces that are impacting an organisation. PEST stands for political, economic, social and technological. Sometimes called PESTEL – referring to political, economic, social, technological, environmental and legal.

Photo call Advance notice to the media of a formally organised opportunity at a set time and date to take a press photograph of a particular person or event.

Place-making A multi-faceted approach to the planning, design and management of public spaces which capitalises on a local community's assets, inspiration and potential, with the intention of creating public spaces that promote people's health, happiness and well-being.

Planning committee The planning decision-making body of a local authority. In the UK, the planning committee is made up of elected members. Its main role is to make decisions on planning applications.

Podcast A digital audio file made available on the internet for downloading to a computer or portable media player.

Practice In architectural terms, a practice is a company providing architectural services.

PQQ (pre-qualification questionnaire) Used to help shortlist suppliers before a tender process.

Project/project team A construction project, sometimes just referred to as a 'project', is the organised process of constructing, renovating, refurbishing, etc a building, structure or infrastructure. The project team is the collection of all organisations and individuals working on the project.

PRCA The Public Relations and Communications Association.

Press release A written communication sent to all news media, usually put out by a representative of a company, organisation or individual.

Press tour Coordinated visits by PR professionals to secure multiple media opportunities.

Pressure group A group of people who work together to try to influence the activities of developers, construction companies, policy makers etc.

Print circulation The total number of copies of a publication available to subscribers as well as via newsstands, vending machines and other delivery systems.

Print production The process of producing printed material such as brochures, posters and leaflets.

Prior approval A procedure whereby permission is deemed granted if the local planning authority does not respond to the developer's application within a certain time.

Project liaison group   A group of stakeholders, usually representative of the wider community, with whom the development team discusses a development proposal on an ongoing basis.

Proptech (property technology)   The use of information technology to help individuals and companies research, buy, sell and manage real estate.

PR panel   See *Colour separations*.

Public affairs   The process of communicating an organisation's point of view on issues or causes to political audiences including MPs and lobbying groups.

Public realm   The external spaces that are accessible to all.

Public sector   All public services in the UK, including emergency services, healthcare, education, social care, housing and refuse collection.

Publics   Target audiences of a company, organisation or individual.

Qualitative research   A scientific method of observation to gather non-numerical data.

Quantitative research   A scientific method of analysing data via statistical, mathematical or computational techniques.

Reach   A data metric that determines the number of people (or percentage of an audience) that have been exposed to content.

Refurbishment   A term used to describe a process of improvement by cleaning, decorating and re-equipping. It may also include elements of retrofitting with the aim of making a building more energy efficient and sustainable.

Regeneration   The use of public money to reverse decline through the transformation of the economic and social geography of a place.

Reputation management   The PR practice of monitoring, correcting and enhancing the perception of a brand, individual, organisation or business in the public's opinion.

Retrofit   The process of installing new or modified parts or equipment in an existing building. Often used in relation to energy efficiency improvements or renewable technologies in existing homes. See also *Refurbishment*.

Return on investment (ROI)   A performance measure used to evaluate the efficiency of an investment.

Retweet   Used in social media: a Twitter user endorses another Twitter user's tweet by forwarding it to their network.

RIBA   Royal Institute of British Architects.

Risk management   Preventive PR whereby an organisation focuses on identifying areas of potential danger and mitigating against a crisis.

RMI   Repair, Maintenance and Improvement sector. Typically referring to home improvements and other smaller building jobs.

ROI   See *Return on investment (ROI)*.

SaaS (Software as a Service)   A way of delivering software licensed on a subscription basis, centrally hosted.

Scattergun   A way of doing something in a way that is not well organised. The scattergun approach in PR is not targeted at particular individuals.

Scheme   See *Project*.

208 *Glossary*

Science-based targets   Science-based targets are a set of goals developed by a business to provide it with a clear route to reduce greenhouse gas emissions. An emissions reduction target is defined as 'science-based' if it is developed in line with the scale of reductions required to keep global warming below 2°C from pre-industrial levels.

Search Engine Optimisation (SEO)   Producing greater visibility for a website by planning and adjusting the content, keywords and phrases of a web page in order to improve its position in search results (search engine ranking).

Sell-in   The process of communicating a news story or idea to a journalist.

Sentiment   The positive or negative feelings and/or perceptions a public has about a given subject.

SEO   See *Search Engine Optimisation (SEO)*.

Share of voice   An advertising revenue model that focuses on weight or percentage as compared to that of other advertisers.

Shares   The number of shares for a social media post or piece of content indicates how many times it has been shared with/passed on to others within social media networks.

Site visit   A visit to a construction site which can provide outside parties, such as media representatives, neighbours, family members of the project team, students and others, with a valuable insight into the work underway on a construction project.

Situational analysis   A collection of methods that can be used to analyse the internal and external environment in relation to a specific proposal. See also PEST analysis and SWOT analysis.

SMART objectives   Goals and objectives that are specific, measurable, achievable, relevant and time-bound.

Snagging   The identification and rectification of faults, defects, mistakes or omissions in a completed construction, whether new or refurbishment, and making them known to the contractor in a snagging list.

Social aggregation sites   Websites that collect content from multiple sources and re-present it in one location.

Social analytics   Search, indexing, semantic analysis and business intelligence technologies used in identifying, tracking, listening to and participating in the distributed conversations about a brand, product or issue, with emphasis on quantifying the trend in each conversation's sentiment and influence.

Social capital   A measure of a community's shared values, relationships and sense of identity.

Social housing   Housing provided for those on low incomes by local authorities, government agencies or non-profit organisations.

Social landlords   Those who own and manage social housing (usually councils or housing associations). Surpluses are re-invested in managing and maintaining existing homes, providing associated services and, in some cases, building new homes.

Social media   Websites and applications that enable users to create and share content or to participate in social networking.

Social responsibility   Providing corporate resources to demonstrate an organisation's commitment to ethically responsible behaviour.

Social value   Social value is created when something improves and contributes to the long-term wellbeing and resilience of individuals, communities and society in general. Public sector bodies often take social value into account through their policies and construction procurement decisions in order to maximise the benefit for the communities they serve.

Soundbite   A short clip of speech, often used to promote or exemplify the full length piece.

Special interest group   A community with a shared interest where members cooperate to affect or to produce solutions in relation to their area of specific interest and may communicate, meet and organise events.

Specification   A specification describes the products, materials and work required by a construction contract.

Specifier   A person who draws up a specification – in construction terms, this could be an architect, designer, contractor or even the client themselves.

Spin   A derogatory term for the act of highlighting the positive aspects of a bad situation, statement or action, usually to the news media.

Spokesperson   A person who is selected and trained to speak on behalf of a company, organisation or brand.

Stakeholder   A person or group with an economic, professional or community interest an organisation's activities.

Stakeholder/publics analysis   The process of analysing stakeholder groups and their (likely) views.

Stakeholder engagement   The process by which an organisation involves people who may be affected by the decisions it makes or can influence the implementation of its decisions.

Stakeholder mapping   The process of identifying stakeholder groups geographically or in other groupings.

Statement of Community Involvement   A document which sets out the processes to be used by the local authority in involving the community in the preparation, alteration and continuing review of all local development documents and development control decisions. The term is also used to refer to the report compiled at the end of a consultation by the development team and submitted as part of the planning application.

Subcontractor   The contractor or trades specialist who is often appointed by the main contractor.

Supply chain   A term used to describe the interconnected hierarchy of supply contracts necessary to procure a built asset. On a 'traditional' building project, a main contractor has a supply chain of sub-contractors and specialist suppliers.

Sustainability   A broad term describing a desire to carry out activities without depleting resources or having harmful impacts, defined by the Brundtland Commission as 'meeting the needs of the present without compromising the ability of future generations to meet their own needs'.

SWOT analysis   An acronym for strengths, weaknesses, opportunities and threats; a structured planning method that evaluates those four elements of a project or business venture.

Syndicated   A news story placed on several websites or in several outlets.

Target audience   The groups to be targeted as recipients of a message.

Third-party certification   The process of getting a company's product, system or process tested and approved by an independent organisation or technical authority.

Thought leader   A recognised authority in a specialised field.

Tier 1 contractor   See *Main contractor*.

Tone   How a person, group, organisation or issue is portrayed in the media; normally categorised as positive, neutral or negative, with various degrees of negative and positive tones.

Topping out   A ceremony traditionally held when the last beam (or its equivalent) is placed atop a structure during its construction; often used as a media event for PR purposes.

Touchables   Technology and devices with touch-screens.

Unique selling point   The specific proposition that a brand or product offers to consumers.

Unique users   Unique IP addresses which have accessed a website. Calculating the number of unique users is a common way of measuring the popularity of a website.

Urban design   The collaborative, multidisciplinary process of shaping the physical setting for life in cities, towns and villages; the art of making places; design in an urban context. It involves the design of buildings, groups of buildings, spaces and landscapes and the establishment of frameworks and processes that facilitate successful development.

User journey   A person's experience during a single website session, consisting of the series of actions performed to reach a specific page.

U-value   A measure of how well an element of a building's fabric – wall, floor, roof, windows, doors and so on – acts as an insulator, ie. how effective they are at preventing heat from transmitting between the inside and the outside of a building. The lower the U-value, the better the thermal performance.

Value engineering   A process designed to solve problems and identify and eliminate unwanted costs, while improving function and quality. The aim is to increase the value of products, satisfying the product's performance requirements at the lowest possible cost. In recent years, it has become synonymous in construction with 'breaking a specification' – the substitution of specified building products with cheaper and inferior alternatives.

Values   Principles important to an organisation, often described alongside its mission and vision.

Video-on-demand   A video media distribution system that allows users to access video entertainment without a traditional video entertainment device and outside of the constraints of a typical static broadcasting schedule.

Virtual reality   A simulated experience that can be similar to or completely different from the real world; can be used for entertainment or educational purposes.

Vision statement   A succinct, realistic, credible, easy-to-understand, relevant and ambitious description of the niche the organisation wants to occupy in the future.

Vlog   A blog created using video content, typically focussed on a cause or special interest.

Webinar   A seminar conducted online.

Wiki   A website combining the ongoing work of many authors, allowing users to modify the content of previous authors.

Wire services   A service for distributing a news release to the media and online. The main wire services are Business Wire, PR Newswire and PRWeb.

Zero carbon   Zero-carbon housing and zero-energy housing are terms used interchangeably to define single homes with a very high energy efficiency ratings and low energy requirements.

Further reading

Cullingworth, B. A. (2006). *Town and Country Planning in the UK*. Routledge.
Curtin, T. (2004). *Managing Green Issues*. London: Palgrave MacMillan.
Department for Communities and Local Government. (2011). *A Plain English Guide to the Localism Act*. Department for Communities and Local Government.
Department for Communities and Local Government. (2014). *The National Planning Policy Framework*. Department for Communities and Local Government.
Doak, G. and Parker, G. (2012). *Key Concepts in Planning*. London: Sage.
Field, J. (August 2017). Stakeholder engagement and advocacy key to infrastructure projects. *Infrastructure Intelligence* (http://www.infrastructure-intelligence.com/article/aug-2017/stakeholder-engagement-and-advocacy-key-infrastructure-projects).
Field, J. (July 2018). Five ways to turn stakeholders into cheerleaders. *CIPR Influence Magazine* Online (https://influenceonline.co.uk/2018/07/23/five-ways-to-turn-stakeholders-into-cheerleaders/).
Field, J. (February 2019). Why better public engagement on transport schemes creates cheerleaders and improves workforce diversity. *Transport Times* (https://www.transporttimes.co.uk/news.php/Why-better-public-engagement-on-transport-schemes-creates-cheerleaders-and-improves-workforce-diversity-365/).
Jones, R. and Gammell, E. (2009). *The Art of Consultation*. London: Biteback Publishing.
Jones, R. and Gammell, E. (2018). *The Politics of Consultation*. Biggleswade: The Consultation Institute.
Norton, P. and Hughes, M. (2018). *Public Consultation and Community Involvement in Planning*. Oxon: Routledge.
Norton, P. and Male, L. (2020). *Promoting Property: Insight, Experience and Best Practice*. Oxon: Routledge.
Rydin, Y. (2011). *The Purpose of Planning*. Bristol: Policy Press.
Wates, N. (2014). *The Community Planning Handbook*. London: Routledge.

Index

Page numbers in **bold** denote tables, in *italic* denote figures

3D: model 23, 26, 29; printing 65, 76, 150

Aarhus Convention 32–33
ABC (Audit Bureau of Circulation) 89, 187
accessibility **17**, 27, 55, 57, 94, 111
accreditation 37, 48, 52, 56, **78**
advertising 63, 86, 89, 91, 124, 137, 145, 152, 169, 188
Advertising Standards Authority (ASA) 169, 171
AEC (architecture, engineering and construction) 147–156, 156n2
agile working 63, 67
analytics 51, 57, 78, 89, 145, 187–188; *see also* Google
approved document **97**
Architects Registration Board (ARB) 39, 41
architecture 3, 39, **44**, 46, 48–51, 56–58, 137, 140; *see also* AEC
artificial intelligence (AI) 57, 65, 150, 157n22, 188–189
Association for Measurement and Evaluation of Communication (AMEC) 145, 187
Association for Project Management 136
audience 14, **16**, 22, 34–36, 39, 41–42, **43**, 44–45, 51–52, 54, 67–72, 93–95, 99–100, 102, 111, 114–115, 120, 128–129, 132–133, 136–137, 139, 141, 145, 166, 177, 188; target **16**, 26, **43**, 46, 75, 80–81, 83, 86–87, 93–95, 99–102, 131, 133–134, 141, 145, 166, 175, 177–178
augmented reality (AR) 23, 65

blockchain 150
blog **15**, 22, 63, 69–72, 82, 85, 87, 154, 158, 161, 179, 188

brand 39, **44**, 56, 67–68, 73, 77, 85, 91–93, 95–96, 98, 124, 138–140, 154, 158, 161, 167–168, 172, 179, 183; ambassador 62, 68–69, 71, 145
BRE Environmental Assessment Method (BREEAM) 81, 98, 140, 146n8, 174
Brexit 14, 61, 67, **78**, 171, 181
Brundtland Commission 159
builders' merchant 91–94, **93**, 101–102
building: consultancies 131–147; control 76–77, **93**
Building Information Modelling (BIM) 7, 50, 65, 76, 148–151, 153–154, 156n12, 161, 185; regulations 76, 80, 95–96, **97**, 98–99, 186; services 3, 159, 167
Built-ID 22, *22–23*
Bus Rapid Transit (BRT) 54
business-to-business (B2B) 8, 50, 87, 96, 101, 168, 178
business-to-consumer (B2C) 8, 175, 189
buying off-plan 46

carbon dioxide (CO_2) 158, 162–163
carbon footprint 162–164
Carillion 64–65
case study 49–50, 52, 63, 79, 83, 85–86, 94–96, 98, 100–102, 111, 138, 142, 178
Chartered Institute of Building (CIOB) 4, 152, 184, 190
Chartered Institute of Marketing (CIM) 55
Chartered Institute of Public Relations (CIPR) 55, 105, 119, 151, 157n22, 171, 177, 189, 191
CHLOE 23–24, *24*
circular economy 81, 160, 164, 178
circulation 89, 152, 187

Climate-Based Daylight Modelling (CBDM) 76
climate change 14, 57, 66, 149, 156, 158–160, 163–164, 173, 175–178
Climate Change Act 163
Code of Conduct 39, 41, 55, 105, 171
colour separations 86, 89, 94
Committee on Climate Change 163
community: benefits 38n1, 46, **47**; engagement 4, 11, 14, 33, 111, 121–123, *125–126*; involvement 11–13, **15**, 33, 37; relations 8, 13
completion 45, **47**, 51–53, **93**, 140
computer-aided design (CAD) 147, 149
computer-generated imagery (CGI) 46, 48, 140
concrete 95, 98, 159, 185; Centre 98; Show 95
Concrete Industry Sustainable Construction Strategy 98
Considerate Constructors Scheme 4
Construction and Property Special Interest Group (CAPSIG) 151
construction industry 1, 3–9, 53, 63, 71, 75, 78, 82–86, 89–90, 92, 148, 151, 161, 165, 179, 182–184, 186–187, 189; *see also* AEC
Construction Industry Training Board (CITB) 62, 183
Construction Leadership Council (CLC) 1, 7–8, 61, 186
Construction Products Association (CPA) 4, 104–105, 182
construction project 1, 6, 185, 188
Construction Sector Deal 6, 150
Construction Skills Network (CSN) 62
construction technology (Contech) 147–158
construction work 3, 4, **43**, 62, 66, 184
consultation: fatigue **15**, 34; online 14, 19, *20–21*, 26–28, 31, 37; public 8, 12, 14, 19, 31, 55, 117, 124, 175; website **18**, 19, *20–21*, *22–23*, 26
ConsultOnline 19, *20–21*
Continuing Professional Development (CPD) 37, 82, 160–161, 191
contractors 1, 3, **5**, 6, **43**, 46, 63–65, 67, 70, **78**, 80, 83, 91, **93**, 96, 110, 121, 137–140, 148, 151; *see also* subcontractors
Corporate Social Responsibility (CSR) 8, 14, 98, 159, 177, 185
Costain Skanska JV (CSJV) 121–122

COVID-19 (coronavirus) 8, 61–62, 65, 67, 82, 95, 107, 137, 152, 154, 156n1, 158, 181, 185–186
crisis communications 45, 89
Crossrail 109, 112, 123–125, *125–126*, 138, 149, 190

data protection **18**, 27
design for manufacture and assembly (DfMA) 150
developer 11–14, **12**, **15**, 19, 22–23, 28, 31, 33, 37, **43**, **47**, 50, 53, 77, **93**, 99, 102, 107–108, 110–112, 115, 132, 137, 152, 164, 172, 174, 182, 184
Development Consent Order (DCO) 108, 112–113
devolved authorities **12**
digital twin 7, 150–151, 157n20, 157n21
Doctrine of Legitimate Expectation 32–33

emissions: carbon 57, 162, 173; direct 162, 164; greenhouse gas 6, 149, 156, 163, 168, 178; indirect 162; low 162
energy efficiency 7, 75, 81, 98, 173, 182, 185
Energy Performance Certificate (EPC) 173
engineering 46, 75–76, 100, 108, 115, 131, 133–134, 137, 139, 141, 147, 156n2, 157n22; *see also* AEC
Environmental, Social & Governance (ESG) 8, 185
Equality and Human Rights Commission 183
Equality Impact Assessment (EqIA) 112, 115
European Union (EU) 61, 171
evaluation 14, **18**, 31, 36–37, 78, 89, 145, 155, 187–189
exclusive 86, 137, 141
Extended Valuation (EV) 27

Facebook 19, 68, 87, 133–134, **135**, 187
fake news 167
Farmer Review 4, 6, 57, 183
fit-out 1, 3, **93**, 103, 174
followers 50, 89–90, 133–134, 136
foundations 3, 29, 48, 58, **93**, 95, 99–101, 182
frameworks 6–7, 14, 24, 41, 69, 79, 104, 131, 145, 156n12, 163–165, 174, 186

General Data Protection Regulation (GDPR) **16**, **18**, 27, 32, 88
geospatial data 26, 150
Glenigan 1, 2

Index   215

Google Analytics 78, 89
greenblush 171
greenhouse gas 158, 162; *see also* emissions
Greenhouse Gas Protocol (GHG Protocol) 162
greenwash 167–169, 171–172
Grenfell Tower 65, 104, 150, 181–182, 185
groundworks **93**, 99–100, 190
Gunning Principles 32, 38n14

Hackitt, J. 104, 150, 182
Hackitt Review 104, 182
hard hats 62, 64
hard to reach groups **15**, 34–35, 123, 185
hashtag (#) 69, 87, 154, 158
High Speed 2 (HS2) 81, 121–122, *122–123*
hoardings 121–122, *122–123*
housing association (HA) 184

impressions 48, 168
industrial strategy 4, 6, 98
influencers 53, 80, 87, 118, 188
infographic 22, 84, *84*
infrastructure 1, 3, 4, **5**, 23, 38n1, 39, 54, 56, 61–62, **97**, 107–115, *110*, 117–118, 120–121, 123, 128–129, 131–132, 136, 138, 142, 146n8, 147, 149, 151, 156n2, 165, 174–175, 177, 185–186
Infrastructure Recovery Programme (IRP) 141
in-house 1, 19, 86, 117, 142, 151
Instagram 50, 68, 71–72, 87, 133–134, **135**, 187–188
Insulslab 99–101
internet 90; non user 27; pre- 67
Internet of Things (IoT) 50, 150

landing page 89–90
Leadership in Energy and Environmental Design (LEED) 81, 174
lean construction 161
legislation 6, **12**, 14, 27, 32, 111
LinkedIn 68, 71–72, 82, 87, 89, 133–134, **135**, 136, 154
lobbying 98, 117
local authority (LA) 12–13, **15**, 24, 30, 35, 38n1, 52, 77, 111–112, 122, 124, 132, 173
local government 8, 55, 107, 151, 182, 186
local planning authority **12**, **16**, 35, **43**
Localism Act (2011) 31, 112

media: earned 188; monitoring 186, 189; owned 179; paid 179; relations 8, 44, 46, 67, 92, 94–95, 98–99, 102–103, 108, 113, 131, 188; social **15**, 19, 26, 28, 45, 50–52, 68, 70–71, **78**, 82–87, 90, 103, 114, 120, 128, 131, 133–134, 136–137, 139, 141–142, 145, 150, 153, 159, 179, 187
messages **16**, 37, 42, 68–69, 71, 94, 98, 102, 104, 109, 113, 117–118, 121, 128, 139–140, 167, 169–170, 179
metrics 145, 148, 189
Minimum Energy Efficiency Standards (MEES) **78**
mixed use 19, 31
Modern Methods of Construction (MMC) 7–8, 76, 86, 96, 150, 185
monitoring 14, **16–17**, 27, 31, 35–36, 89, 150; *see also* media
Mott MacDonald 131, 133–134, **135**, 136, 141–142, 165

National Infrastructure Commission's Data for the Public Good 150
National Planning Policy Framework (NPPF) 12–13, **12**, 31
Nationally Significant Infrastructure Project (NSIP) 12, **12**, 38n1, 107, 111–112, 114
neighbourhood forum **12**, 184
net zero **43**, 160, 163–164

O&H Properties 19
offsite construction 161; *see also* Modern Methods of Construction (MMC)
One Planet Living 165, 174

parish council **12**, **15**
Passivhaus 76, 81, 86
Pershore Town Plan 29–30, *30*
PESO (Paid, Earned, Shared and Owned) model 91, 105n1, 177, 179, 189
plan 13, **17**, 44, 46, 55, 61, 70, 89, 122, 131, 145, 165, 176, 191; action 30, 174, 183; district; local **15**, **43**; master- 30; neighbourhood **15**; strategic 30; town 30
planning: applications 12–13, **12**, **15–16**, **43**, 111; pre- 13, 112; authority 13, 174; committee **15**; planning consent **47**, 50, 112; Inspectorate 107, 112; permission **12**, 37, 42, 45, 52, 119; for Real 29–30; *see also* local planning authority
podcast 22, 69, 71, 87, 136, 153, 188
policy makers 112, 126, 175
power and interest grid 113–114, *114*
practice: best 4, 37, 56, 69, 95–96, 111–113, 116–117, 121, 123, 127, 136, 161; good 83, 122, 160, 162, 183; manager 9; working 9, 67

pre-qualification questionnaire (PQQ) 42
press release **18**, 46, 83, 94, 128, 132, 137–140, 145, 188, 191
pressure group 26, 34
project: management 76, **93**, 131; team 3, 45, 128, 132, *143–144*, 147–148, 156n12
property technology (Proptech) 24, 157n21
prototyping 156n12
public affairs 108, 113, 127
public inquiry 181
public meeting 28, 120
Public Relations and Communications Association (PRCA) 151
publics 14, 116, 178

qualitative research **17–18**, 31
quantitative research **17–18**, 31, 128

recycling 66, 160
refurbishment 1, 6, 39, **43**, 45, **78**, 83, 151, 164, 174–175, 181
regeneration 3, 29, **43**
retrofit 112, 115, 182; *see also* refurbishment
return on investment (ROI) 133, 145, 189, 191
Royal Institute of British Architects (RIBA) 4, 39, **43**, 50, 57, 82, 86; Journal 42, 86; Plan of Works 39, *40*
Royal Institution of Chartered Surveyors (RICS) 4, 136, 174

science-based targets 172–173
search engine optimisation (SEO) 188
Secure Sockets Layer (SSL) 27
sentiment 13, 29, 31, 115, 128, 159
site visit 140–142
situational analysis 14, **15**
SMART objectives (specific, measurable, achievable, relevant and time-bound) 128, 132, 178
social: network 158; value 49, 52–53, 66, 81, 110–111, 160–161; *see also* media
Social Value Act 6, 53, 66, 110
Software as a Service (SaaS) 150
special interest group **15**, 151, 153
specification 1, 91–93, 96, 98–99, 102, 182
specifier 86–87, 91, 96, 99–100, 102, 104
stakeholder: analysis **15**, 35; engagement 8, 49, 53, 54, 108, 113, 117, 120, 126, 128–129; groups 36, 114, 117–118, 128, 166; mapping **15**, 113
Statement of Community Involvement 13, **15**
subcontractors 3, **5**, 75–78, 81, 83, 85–86, 90–92, **93**, 96
supply chain 5, 8, 9, 41, 62, 80, 91–92, 96, 101, 103–104, 131, 151, 159–160, 162, 164, 167, 173, 179, 187, 189
Supply Chain Sustainability School 160
sustainability 3, 52, 63, 72, 81, **93**, 99, 131, 140, 146n8, 159–161, 165–167, 171–179, 185
Sustainable Development Goals (SDG) 164–166, 178
SWOT (strengths, weaknesses, opportunities and threats) 4, **5**, 38n6, 77, **78**

thermal performance 99–100
third-party certification 168, 172–173
thought leader 44, 46, 67, 81, 85–87, 92, 98–99, 101, 136, 175, 190
tone of voice 68, 85
Trade Fabrication Systems (TFS) 101–104
trade publications 58, 84, 136
tradespeople **78**, 182
Transport for London (TfL) 116, 124–127
Twitter 19, 50, 68, 72, 87, 133–134, **135**, 154, 187

UK construction 3–4, **5**, 9, 149, 181
UK Construction Labour Model 4, 57, 183
UK Green Building Council (UKGBC) 53, 160–164, 178
urban design 39, 54
U-value 100

value engineering 46, 75
values 65, 67–69, 122, 167, 175
virtual reality (VR) 26, 150
vlog 22, 70–71
VU.CITY 24–25, *25*

webinar 68–69, 71, 145, 153
WELL Building Standard 81, 174
Wikipedia 64, 167
Willmott Dixon 63–65, 68–70

YouTube 1, 19, 50, 63, 68, 71, 87, 133–134, **135**, 154, 190

zero carbon 98, 163–164